SOLVING ALGEBRA WORD PROBLEMS

Judy Barclay

Cuesta College

THOMSON

BROOKS/COLE

Australia • Canada • Mexico • Singapore • Spain
United Kingdom • United States

THOMSON
™
WADSWORTH

Acquisitions Editor: Jennifer Huber
Assistant Editor: Rebecca Subity
Technology Project Manager: Christopher Delgado
Marketing Manager: Leah Thomson
Advertising Project Manager: Bryan Vann
Project Editor: Belinda Krohmer
Art Director: Vernon Boes

Print/Media Buyer: Rebecca Cross
Permissions Editor: Sommy Ko
Copy Editor: Carol Reitz
Cover Designer: Andy Norris
Cover Image: Bob Anderson/Masterfile
Cover Printer: Webcom Limited
Compositor: G & S Typesetters
Printer: Webcom

For more information
about our products, contact us at:
**Thomson Learning Academic
Resource Center
1-800-423-0563**

For permission to use material from this text,
contact us by: **Phone:** 1-800-730-2214
Fax: 1-800-730-2215
Web: http://www.thomsonrights.com

Library of Congress Control Number:
2003114069

ISBN-13: 978-0-534-49573-2
ISBN-10: 0-534-49573-7

Brooks/Cole—Thomson Learning
10 Davis Drive
Belmont, CA 94002
USA

Asia
Thomson Learning
5 Shenton Way #01-01
UIC Building
Singapore 068808

Australia/New Zealand
Thomson Learning
102 Dodds Street
Southbank, Victoria 3006
Australia

Canada
Nelson
1120 Birchmount Road
Toronto, Ontario M1K 5G4
Canada

Europe/Middle East/Africa
Thomson Learning
High Holborn House
50/51 Bedford Row
London WC1R 4LR
United Kingdom

Latin America
Thomson Learning
Seneca, 53
Colonia Polanco
11560 Mexico D.F.
Mexico

Spain/Portugal
Paraninfo
Calle Magallanes, 25
28015 Madrid, Spain

TABLE OF CONTENTS

To the Student

In my thirty years of teaching algebra, I have encountered thousands of students who felt they would never be able to solve word problems. Word problems can be frightening and can hinder a student's ability to succeed in elementary and intermediate algebra courses. Many students will just skip all the word problems in the assignment and on the test. I always tell my students that they have to be like the little engine that could; they have to think positively and try the problems. If you never try the problems, you definitely will never be able to solve them.

Problem solving is a skill that is vital to success in all mathematics courses and is also important in physics, statistics, and other disciplines. The purpose of this book is to help you learn how to solve word problems algebraically and to take some of the frustration out of learning this skill.

My approach in this book is to give you a five-step strategy to follow in solving word problems. I use these steps to solve many examples in each chapter. Then I provide similar exercises for you to work on. All the answers are given at the end of the book. I suggest that you work all the examples with pencil and paper first. After you thoroughly understand the examples in each chapter, you should practice by doing the exercises at the end of the chapter.

There is no substitute for working the problems on your own, even if you have to struggle a little with some of them. With lots of practice solving these problems, you will increase your confidence, which in turn will lower your anxiety. You wouldn't expect to become a good golfer or tennis player without lots of practice. The same is true in mathematics. By developing this important skill, you will become a better problem solver and will be able to apply these skills to problems in other courses.

When you finish your course, keep this book. It will be a good reference for other courses in the future. I hope this step-by-step approach will help you to succeed in your algebra course. Good luck and happy problem solving!

To the Instructor

Solving Algebra Word Problems includes all the most common types of word problems encountered in elementary and intermediate algebra textbooks. I have made an effort to make the problems in the examples and exercises interesting and relevant.

The emphasis is on reasoning out the solution using a five-step strategy. These steps are used on every example in the book. This strategy should be consistent with whatever textbook you are using for your algebra course.

I have designed this book to be used as a supplement to your regular algebra textbook or as the primary text for a short course on problem solving. It can be used in a classroom setting or by the student individually with a tutor.

Acknowledgments

I would like to thank all those who assisted me in the development and completion of this book: the many students I have had in classes over the past thirty years, my colleagues, my friends and family, and the staff at Thomson Learning.

I want to thank Pat McKeague for his suggestions, encouragement, and support of this project. His help was invaluable. Thanks also go to my dean, Ann Grant, and the math faculty at Cuesta College who were supportive of my desire to write this book. I am grateful to Jen Sanders for her suggestions and comments.

Thanks to Jennifer Huber who had confidence in my project. Jennifer, Belinda Krohmer and the staff at Thomson Learning did an excellent job producing this book.

Special thanks to my daughters, Stephanie and Erika, and to my husband, Ken, for their encouragement and input into this project. Ken wishes he had this word problem book when he was taking algebra.

From English to Algebra

The first skill required for solving word problems is the ability to translate English statements into mathematical symbols. We will begin by practicing this skill. Below are some examples of translations from English to algebra, grouped by the mathematical operation that is required.

Examples Using Addition

1. Write the following English phrases as algebraic expressions. In each case, let x = the number or the unknown.

English Phrase	Algebraic Expression
a. The sum of a number and five	$x + 5$
b. Four more than a number	$4 + x$
c. Ten increased by a number	$10 + x$
d. The sum of the unknown and fifteen	$x + 15$
e. The total of three and a number	$3 + x$

2. Write the following in algebraic notation, stating what x represents: Helen's age four years from now.

   Solution   Let x = Helen's age now.
   Helen's age in four years = $x + 4$.

Examples Using Subtraction

3. Write the following English phrases as algebraic expressions. In each case, let x = the number or the unknown.

English Phrase	Algebraic Expression
a. Two less than the unknown	$x - 2$
b. Seven decreased by a number	$7 - x$
c. The difference of a number and eleven	$x - 11$
d. A number minus three	$x - 3$
e. Subtract a number from twelve	$12 - x$

4. Write the following in algebraic notation, stating what x represents: Bob's age eight years ago.

Solution   Let x = Bob's age now.
Bob's age eight years ago = $x - 8$.

Examples Using Multiplication

5. Write the following in algebraic notation. In each case, let x = the number or the unknown.

English Phrase	Algebraic Expression
a. Twice the unknown	$2x$
b. Six times a number	$6x$
c. The product of seven and a number	$7x$
d. Fifteen percent of a number	$0.15x$
e. One-fifth of the unknown	$\dfrac{1}{5}x$

6. Write the following in algebraic notation, stating what x represents:
   a. The value of x dimes in cents
   b. The distance traveled in x hours at 40 miles per hour

Solution
   a. Let x = the number of dimes. Because each dime is worth 10 cents, the value of the dimes is $10x$.
   b. Let x = the number of hours. Because distance equals rate multiplied by time, or $d = r \cdot t$, the distance traveled is $40x$.

Examples Using Division

7. Write the following as algebraic expressions. In each case, let $x =$ the number.

English Phrase	Algebraic Expression
a. The quotient of a number and eight	$\dfrac{x}{8}$
b. Four divided by a number	$\dfrac{4}{x}$
c. The quotient of eleven and a number	$\dfrac{11}{x}$
d. The ratio of a number and six	$\dfrac{x}{6}$

Other Examples

8. Write the following in algebraic notation, stating what x represents:
   a. The reciprocal of a number
   b. The opposite of a number
   c. The sum of two consecutive integers
   d. The sum of two consecutive even integers
   e. The sum of three consecutive odd integers
   f. Four more than twice a number
   g. Five less than three times a number

Solution

a. Let $x =$ the number. The reciprocal of the number is $\dfrac{1}{x}$.

b. Let $x =$ the number. The opposite of the number is $-x$.

c. Let $x =$ the first consecutive integer. Then the second consecutive integer is $x + 1$. The sum of these two consecutive integers is $x + x + 1$, or $2x + 1$.

d. Let $x =$ the first consecutive even integer. Then the second consecutive even integer is $x + 2$. The sum of these two consecutive even integers is $x + x + 2$, or $2x + 2$.

   e. Let x = the first consecutive odd integer. Then the second consecutive odd integer is $x + 2$ and the third is $x + 4$. The sum of these three consecutive odd integers is $x + x + 2 + x + 4$, or $3x + 6$.

   f. Let x = the number. Four more than twice the number is $4 + 2x$.

   g. Let x = the number. Five less than three times the number is $3x - 5$.

The next skill that we will practice is translating English statements into algebraic equations. The words *is*, *was*, and *will be* become the equal sign (=). You will notice that English phrases translate into algebraic expressions and English statements translate into algebraic equations. A statement has a verb and a phrase does not.

Examples Involving Translating Equations

9. Translate the following statements into algebraic equations. In each one, let x = the number.

English Statement	**Algebraic Equation**
a. The sum of a number and three is 200.	$x + 3 = 200$
b. The difference of fifteen and a number is -3.	$15 - x = -3$
c. Three more than five times a number is twelve.	$3 + 5x = 12$
d. Thirty percent of a number is forty.	$0.30x = 40$
e. Three more than twice the reciprocal of a number is $\dfrac{13}{3}$.	$3 + 2\left(\dfrac{1}{x}\right) = \dfrac{13}{3}$
f. Five times the difference of twice a number and six equals twenty.	$5(2x - 6) = 20$

10. Write an equation for the following statement, stating what x represents: The sum of two consecutive integers is 37.

   Solution   Let x = the first consecutive integer.
                Then the next consecutive integer is $x + 1$.
                The equation is $x + x + 1 = 37$.

PROBLEM SET 1

In each of the following problems, translate the English phrase into an algebraic expression. Let x = the number.

1. The sum of a number and four

2. The sum of eight and a number

3. Five more than a number

4. Seven more than a number

5. Six increased by a number

6. A number increased by eight

7. The total of nine and a number

8. The total of two and a number

9. Three less than a number

10. Five less than a number

11. Eight decreased by a number

12. Twelve decreased by a number

13. The difference of a number and six

14. The difference of eleven and a number

15. Subtract thirteen from a number

16. Subtract a number from twenty

17. Three times a number

18. Ten times a number

19. The product of a number and six

20. The product of twelve and a number

21. Fifty percent of a number

22. Twenty percent of a number

23. Three-fourths of a number

24. Two-thirds of a number

25. The quotient of four and a number

26. The quotient of a number and seven

27. Ten divided by a number

28. A number divided by eight

29. The ratio of a number and five

30. The ratio of twenty and a number

31. Three more than twice a number

32. Six more than ten times a number

33. Two less than three times a number

34. Five less than six times a number

35. The reciprocal of twice a number

36. The reciprocal of two more than a number

37. Twice the sum of a number and three

38. Three times the sum of four and a number

39. Ten times the difference of a number and twelve

40. Five times the difference of eight and a number

Write an English phrase for each of the following algebraic expressions. Note: Answers may vary.

41. $x + 6$

42. $8 + x$

43. $12 - x$

44. $x - 10$

45. $2x + 5$

46. $3x + 9$

47. $4x - 3$

48. $5x - 2$

49. $4(3x - 6)$

50. $5(8x + 3)$

51. $\dfrac{x}{9}$

52. $\dfrac{1}{x - 3}$

Write an algebraic equation for each of the following English statements. In each case, state what x represents.

53. The sum of a number and seven is thirty-two.

54. The sum of a number and twelve is 102.

55. The difference of a number and three is twenty-two.

56. The difference of eight and a number is forty-one.

57. If two-thirds of a number is increased by seven, the result is seventeen.

58. If three-fourths of a number is decreased by five, the result is four.

59. If you subtract three from a number, the result is nineteen.

60. If you subtract thirteen from a number, the result is fifty-two.

61. Four more than twice a number is thirty-four.

62. Five more than three times a number is twenty-six.

63. The sum of two consecutive even integers is seventy-two.

64. The sum of two consecutive odd integers is eighty-two.

65. Two more than three times the reciprocal of a number is seven-thirds.

66. Three less than five times the reciprocal of a number is negative eight-thirds.

67. Seven percent of a number is ninety-eight.

68. Twelve percent of a number is 168.

2

Integer Problems

The key to successful problem solving is practice. You must practice different types of word problems in order to become a good problem solver. Just as a tennis player or a golfer must practice to be good at the sport, you must practice solving problems to become good at it.

Algebra is the tool we use to solve problems. Some of these problems can be solved without algebra. However, I encourage you to solve them all algebraically. We practice with the easier problems so that we can succeed with the harder ones.

You must analyze the problem, devise a plan, and then carry out your plan. The following steps are the strategy we will use to solve problems in this book.

Strategy For Solving Word Problems

1. <u>Read</u> the problem <u>and write</u> what is <u>given and</u> what you are asked to find (<u>the unknown</u>). It is also helpful to decide what type of problem it is.
2. <u>Begin</u> the solution <u>with</u> "<u>Let</u> x = the unknown." Represent other unknowns in terms of x. If possible, make a diagram or chart that relates the given and unknown quantities.
3. Go back, <u>reread</u> the problem, <u>and write an equation</u> that relates the given quantities.
4. <u>Solve the equation</u> and write your answer in a sentence using appropriate labels.

5. <u>Check</u> your answer back into the original statement of the problem.

We will now use these steps to solve integer problems.

EXAMPLE 1 **The larger of two integers is one less than twice the smaller integer. Their sum is 32. Find the two integers.**

Solution

Step 1 Read and write givens and unknowns.

We want to find two integers. We are given that their sum is 32 and that the larger is one less than two times the smaller.

Step 2 Begin with "Let x = the unknown."

Let x = the smaller integer
$2x - 1$ = the larger integer

Step 3 Reread and write an equation.

Their sum $= 32$
$x + 2x - 1 = 32$

Step 4 Solve the equation.

$$x + 2x - 1 = 32$$
$$3x - 1 = 32 \qquad\qquad 2x - 1 = 2(11) - 1$$
$$3x = 33 \qquad\qquad\qquad = 22 - 1$$
$$x = 11 \qquad\qquad\qquad = 21$$

The two integers are 11 and 21.

Step 5 Check.

The sum of 11 and 21 is 32, and 21 is one less than twice 11. It checks.

EXAMPLE 2 **Three more than twice an integer is equal to six less than five times the same integer. Find the integer.**

Solution

Step 1 Read and write givens and unknowns.

We want to find an integer. We are given two ways to represent the same integer: (1) three more than two times the integer and (2) six less than five times the integer.

Step 2   Begin with "Let x = the unknown."

Let x = the integer

Step 3   Reread and write an equation.

Since we have two ways to represent the same integer, these two are equal:

3 more than 2 times the integer = 6 less than 5 times the integer

$$3 + 2x = 5x - 6$$

Step 4   Solve the equation.

$$3 + 2x = 5x - 6$$
$$2x = 5x - 9$$
$$-3x = -9$$
$$x = 3$$

The integer is 3.

Step 5   Check.

3 more than 2 times 3 is $2(3) + 3 = 6 + 3 = 9$.
6 less than 5 times 3 is $5(3) - 6 = 15 - 6 = 9$. It checks.

EXAMPLE 3   **There are three consecutive even integers. Twice the first is 10 more than the third. Find the integers.**

Solution

Step 1   Read and write givens and unknowns.

We are looking for three consecutive even integers. We are given that two times the first integer equals 10 added to the third integer. The problem doesn't tell us anything about the second integer, so we won't use it in the equation.

Step 2   Begin with "Let x = the unknown."

Let x = the first consecutive even integer
$x + 2$ = the second consecutive even integer
$x + 4$ = the third consecutive even integer

Step 3   Reread and write an equation.

2 times the first = 10 added to the third

$$2 \cdot x = 10 + x + 4$$

Step 4 Solve the equation.

$$2x = 10 + x + 4 \qquad x + 2 = 14 + 2 = 16$$
$$2x = 14 + x$$
$$x = 14 \qquad\qquad x + 4 = 14 + 4 = 18$$

The three consecutive even integers are 14, 16, and 18.

Step 5 Check.

2 times the first integer is $2(14) = 28$.
10 added to the third integer is $10 + 18 = 28$. It checks.

EXAMPLE 4 **A 37-foot log is cut into two pieces. The longer piece is 1 foot more than twice the shorter piece. Find the length of each piece.**

There are two ways to solve this problem. We will do it both ways.

Solution

Step 1 Read and write givens and unknowns.

We want to find the length of each piece. We are given that the total length is 37. The longer piece is twice the shorter piece plus one.

Step 2 Begin with "Let x = the unknown."

Let x = the shorter piece
$2x + 1$ = the longer piece

Step 3 Reread and write an equation.

The total or sum = 37
$$x + 2x + 1 = 37$$

Step 4 Solve the equation.

$$x + 2x + 1 = 37$$
$$3x + 1 = 37 \qquad\qquad 2x + 1 = 2(12) + 1$$
$$3x = 36 \qquad\qquad\qquad = 24 + 1$$
$$x = 12 \qquad\qquad\qquad = 25$$

The two pieces of the log are 12 feet and 25 feet long.

Step 5 Check.

The sum of 12 and 25 is 37, and 25 is one more than twice 12. It checks.

Alternate Solution

Step 1 Read and write givens and unknowns.

We want to find the two lengths. We are given that the total length is 37. The longer piece is twice the shorter piece plus one.

Step 2 Begin with "Let x = the unknown."

Because the sum of the two pieces is 37, we will let the length of one piece be x and the other will be the total minus x, or $37 - x$, because $x + (37 - x) = 37$.

Let x = the shorter piece
$37 - x$ = the longer piece

Step 3 Reread and write an equation.

The longer piece = twice the shorter piece plus one
$$37 - x = 2 \cdot x + 1$$

Step 4 Solve the equation.

$$
\begin{aligned}
37 - x &= 2x + 1 \\
-x &= 2x - 36 \\
-3x &= -36 \\
x &= 12
\end{aligned}
\qquad\qquad
\begin{aligned}
37 - x &= 37 - 12 \\
&= 25
\end{aligned}
$$

The two pieces of the log are 12 feet and 25 feet long.

Step 5 Check.

The sum of 12 and 25 is 37, and 25 is one more than twice 12. It checks.

EXAMPLE 5 **The sum of two integers is 25. One less than twice the larger integer is equal to one more than four times the smaller integer. Find the integers.**

Solution

Step 1 Read and write givens and unknowns.

We are looking for two integers whose sum is 25. We are given that two times the larger integer minus one equals four times the smaller integer plus one.

Step 2 Begin with "Let x = the unknown."

Because the sum of the two integers is 25, we will let one of the integers be x and the other will be $25 - x$.

Let x = the larger integer

$25 - x$ = the smaller integer

Step 3 Reread and write an equation.

2 times the larger integer minus 1 = 4 times the smaller integer plus 1

$$2 \cdot x - 1 = 4 \cdot (25 - x) + 1$$

Step 4 Solve the equation.

$$2x - 1 = 4(25 - x) + 1$$
$$2x - 1 = 100 - 4x + 1$$
$$2x - 1 = 101 - 4x$$
$$2x = 102 - 4x$$

$$6x = 102 \qquad\qquad 25 - x = 25 - 17$$
$$x = 17 \qquad\qquad\qquad = 8$$

The larger integer is 17 and the smaller integer is 8.

Step 5 Check.

The sum of 17 and 8 is 25. Twice the larger minus one = $2(17) - 1 = 34 - 1 = 33$. Four times the smaller plus one = $4(8) + 1 = 32 + 1 = 33$. It checks.

EXAMPLE 6 **Three more than twice the reciprocal of an integer is equal to $\dfrac{11}{3}$. Find the integer.**

Solution

Step 1 Read and write givens and unknowns.

We want to find an integer. If three is added to two times the reciprocal of the integer, the result is $\dfrac{11}{3}$. The reciprocal of an integer is 1 divided by that number.

Step 2 Begin with "Let x = the unknown."

Let x = the integer

$\dfrac{1}{x}$ = the reciprocal of the integer

Step 3 Reread and write an equation.

$$3 \text{ added to 2 times the reciprocal of the integer} = \frac{11}{3}$$

$$3 + 2 \cdot \frac{1}{x} = \frac{11}{3}$$

Step 4 Solve the equation.

$$3 + 2\left(\frac{1}{x}\right) = \frac{11}{3}$$

$$3 + \frac{2}{x} = \frac{11}{3}$$

$$3x(3) + 3x\left(\frac{2}{x}\right) = 3x\left(\frac{11}{3}\right) \qquad \text{Multiply each side by } 3x.$$

$$9x + 6 = 11x \qquad \text{Simplify each side.}$$

$$6 = 2x$$

$$3 = x$$

The integer is 3.

Step 5 Check.

$$3 \text{ added to twice the reciprocal of } 3 \text{ is } 3 + 2\left(\frac{1}{3}\right) = 3 + \frac{2}{3} = \frac{9}{3} + \frac{2}{3} = \frac{11}{3}.$$

It checks.

PROBLEM SET 2

Solve each of the following integer problems. Be sure to follow the strategy for solving word problems presented at the beginning of this chapter.

1. The larger of two integers is seven more than the smaller integer. Their sum is 49. Find the integers.

2. The larger of two integers is one more than twice the smaller integer. Their sum is 43. Find the integers.

3. The larger of two integers is two more than three times the smaller integer. Their sum is 70. Find the integers.

4. The larger of two integers is one less than four times the smaller integer. Their sum is 74. Find the integers.

5. Seven less than three times an integer is equal to eight more than twice the integer. Find the integer.

6. Five less than four times the integer is equal to seven more than three times the integer. Find the integer.

7. Six more than three times the integer is equal to eight less than five times the integer. Find the integer.

8. Twice an integer increased by 15 is equal to three times the integer decreased by five. Find the integer.

9. The sum of three consecutive odd integers is 69. Find the integers.

10. The sum of three consecutive integers is 36. Find the integers.

11. There are three consecutive odd integers. Twice the first integer is 25 more than the third. Find the integers.

12. There are three consecutive even integers. The sum of the first two integers is 16 more than the third. Find the integers.

13. A 24-foot board is cut into two pieces. The longer piece is 3 feet more than twice the shorter piece. Find the length of each piece.

14. A 60-inch submarine sandwich is cut into three pieces. The longest piece is twice the shortest piece, and the middle piece is 4 inches more than the shortest piece. Find the length of each piece.

15. Three volunteers work together stuffing 90 envelopes. Susan stuffs twice as many envelopes as Judy, and Valerie stuffs three times as many envelopes as Judy. How many envelopes did each of the volunteers stuff?

16. A 45-centimeter pipe is cut into two pieces. The longer piece is 3 centimeters more than twice the shorter piece. Find the lengths of the two pieces.

17. The sum of two integers is 45. One more than three times the smaller is equal to four less than twice the larger. Find the integers.

18. The sum of two integers is 66. One more than four times the smaller is equal to five less than twice the larger. Find the integers.

19. The sum of two integers is 33. Two more than three times the smaller is equal to one more than twice the larger. Find the integers.

20. The sum of two integers is 81. Three less than twice the larger is equal to seven more than twice the smaller. Find the integers.

21. Two more than three times the reciprocal of an integer is $\dfrac{9}{4}$. Find the integer.

22. Five minus twice the reciprocal of an integer is $\dfrac{39}{8}$. Find the integer.

23. Three minus four times the reciprocal of an integer is $\dfrac{23}{8}$. Find the integer.

24. Three more than six times the reciprocal of an integer is $\dfrac{16}{5}$. Find the integer.

25. Five times the difference of eight and a number is 25. Find the number.

26. Four times the difference of a number and 12 is 52. Find the number.

27. Seven times the sum of a number and four is 112. Find the number.

28. Three times the sum of a number and seven is 48. Find the number.

29. According to the U.S. Department of Defense, women made up about 15% of the U.S. military in 2002. There were a total of 210,177 women in all branches of the military. The number of women in the Navy was 54,677 and the number in the Air Force was 70,743. The number in the Army was 1109 more than seven times the number in the Marines. Find the number of women in the Marines.

30. Use of the Internet among Americans is growing. According to "A Nation On-line: How Americans Are Expanding Their Use of the Internet" compiled from the 2001 U.S. Census Bureau's population survey, the states of Washington and Oregon had a total of 3671 households with Internet access. Washington had 1019 more households than Oregon did. Find the number of households with Internet access in these two states.

The following problems require algebraic techniques from intermediate algebra to solve:

31. The sum of two positive integers is 20 and their product is 75. Find the integers.

32. The sum of two positive integers is 21 and their product is 108. Find the integers.

33. The sum of the reciprocals of two positive integers is $\frac{1}{4}$. One of the integers is twice the other. Find the integers.

34. The difference of the reciprocals of two positive consecutive even integers is $\frac{1}{60}$. Find the integers.

35. The sum of the squares of two consecutive odd integers is 130. Find the integers. (There are two answers.)

36. The sum of the squares of two consecutive integers is 145. Find the integers. (There are two answers.)

37. During the 2002 regular season of the National Football League, the league leaders in passing were Rich Gannon and Drew Bledson. The total of their passing yards was 9048. The difference in total passing yards between Gannon and Bledson was 330 yards. Find the number of yards each one passed for during the 2002 season.

Geometry Problems

In this chapter we will practice solving geometry problems with the same strategy we used in Chapter 2. Here are some extra tips and formulas for solving geometry problems.

Tips

1. Memorize the formulas for the area and perimeter of geometric figures listed below.
2. Draw a picture of the figure and label the sides (or angles).
3. Label the answer with the appropriate units.
4. Make certain that all units are the same. If not, change them so that they all agree.

Formulas to Remember

Rectangle: Perimeter = twice the length plus twice the width or
$P = 2l + 2w$
Area = length times width or $A = l \cdot w$

Square: Perimeter = four times the side or $P = 4s$
Area = the square of the side or $A = s^2$

Circle: Circumference = 2π times the radius or $C = 2\pi r$
Area = π times the square of the radius or $A = \pi r^2$

Triangle: Perimeter = sum of the sides or $P = a + b + c$
Area = one-half the base times the height or

$$A = \frac{1}{2}bh$$

The sum of the interior angles of a triangle equals $180°$.
Complementary angles are two angles whose sum is $90°$.
Supplementary angles are two angles whose sum is $180°$.
In a right triangle, $c^2 = a^2 + b^2$, where c is the hypotenuse and a and b are the legs.

We will now solve some geometry problems.

EXAMPLE 1   **The width of a rectangle is twice the length. The perimeter is 42 centimeters. Find the dimensions of the rectangle.**

Solution

Step 1   Read and write givens and unknowns.

We want to find the length and width of a rectangle. We know that its width is twice its length and the perimeter is 42 cm.

Step 2   Begin with "Let x = the unknown."

Let x = the length
    $2x$ = the width

Step 3   Reread and write an equation.

The perimeter is 42 cm.
Twice the length plus twice the width = 42
$$2(x) + 2(2x) = 42$$

Step 4   Solve the equation.

$$2x + 2(2x) = 42$$
$$2x + 4x = 42$$
$$6x = 42 \qquad 2x = 2(7)$$
$$x = 7 \qquad = 14$$

The length of the rectangle is 7 cm and the width is 14 cm.

Step 5   Check.

The width is twice the length. The perimeter is $2(7) + 2(14) = 14 + 28$ or 42. It checks.

EXAMPLE 2 The first side of a triangle is twice the second side. The third side is three times the second side. If the perimeter of the triangle is 210 feet, find the three sides of the triangle.

Solution

Step 1 Read and write givens and unknowns.

We want to find the three sides of a triangle. We know that the first side is two times the second side and the third side is three times the second side. Also, the perimeter is 210 ft.

Step 2 Begin with "Let x = the unknown."

Let x = the second side
$2x$ = the first side
$3x$ = the third side

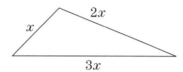

Step 3 Reread and write an equation.

The perimeter or the sum of the sides = 210
$$x + 2x + 3x = 210$$

Step 4 Solve the equation.

$$x + 2x + 3x = 210$$
$$6x = 210 \qquad\qquad 2x = 2(35) = 70$$
$$x = 35 \qquad\qquad 3x = 3(35) = 105$$

The three sides of the triangle are 35 ft, 70 ft, and 105 ft.

Step 5 Check.

The perimeter of the triangle is $35 + 70 + 105$, or 210 ft. Also, 70 is twice 35 and 105 is three times 35. It checks.

EXAMPLE 3 In triangle ABC, the measure of angle B is twice the measure of angle A and the measure of angle C is five more than four times the measure of angle A. Find the measures of the three angles of the triangle.

Solution

Step 1 Read and write givens and unknowns.

We want to find the measures of the three angles of $\triangle ABC$. We know that $\angle B$ is two times $\angle A$ and $\angle C$ is five more than four times $\angle A$. The sum of the interior angles of a triangle is always $180°$.

Step 2   Begin with "Let x = the unknown."

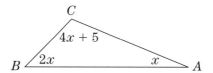

Let x = the degree measure of $\angle A$
$2x$ = the degree measure of $\angle B$
$4x + 5$ = the degree measure of $\angle C$

Step 3   Reread and write an equation.

The sum of the angles = 180°
$$x + 2x + 4x + 5 = 180$$

Step 4   Solve the equation.

$$x + 2x + 4x + 5 = 180$$
$$7x + 5 = 180$$
$$7x = 175 \qquad\qquad 2x = 2(25) = 50$$
$$x = 25 \qquad\qquad 4x + 5 = 4(25) + 5 = 105$$

The angles are 25°, 50°, and 105°.

Step 5   Check.

The sum of the angles is 25° + 50° + 105°, or 180°.
Also, 50° is 2 times 25° and 105° is 4 times 25° plus 5°. It checks.

EXAMPLE 4   **Angle ABC and angle DBC are supplementary. The measure of angle ABC is three more than twice the measure of angle DBC. Find the measures of the two angles.**

Solution

Step 1   Read and write givens and unknowns.

We want to find the measures of the two angles. We know that $\angle ABC$ is 3 added to 2 times $\angle DBC$ and that the sum of the two angles is 180° (because they are supplementary angles).

Step 2   Begin with "Let x = the unknown."

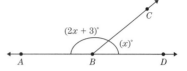

Let x = the degree measure of $\angle DBC$

$2x + 3$ = the degree measure of $\angle ABC$

Step 3   Reread and write an equation.

The sum of the two angles = $180°$

$$x + 2x + 3 = 180$$

Step 4   Solve the equation.

$$x + 2x + 3 = 180$$
$$3x + 3 = 180$$
$$3x = 177$$
$$x = 59 \qquad\qquad 2x + 3 = 2(59) + 3 = 121$$

Angle ABC is $121°$ and angle DBC is $59°$.

Step 5   Check.

The sum of $121°$ and $59°$ is $180°$. Also, $121° = 2(59°) + 3°$. It checks.

EXAMPLE 5   **A vegetable garden is in the shape of a right triangle. The hypotenuse is 1 foot more than twice the shortest side. The other side is 1 foot less than twice the shortest side. Find the length of the shortest side.**

Solution

Step 1   Read and write givens and unknowns.

We want to find the length of the shortest side of the right triangle. We know that the hypotenuse is 1 foot more than twice the shortest side. The second side is 1 foot less than twice the shortest side. We also know that in a right triangle $c^2 = a^2 + b^2$, where c is the hypotenuse.

Step 2   Begin with "Let x = the unknown."

Let x = the shortest side

$2x + 1$ = the hypotenuse

$2x - 1$ = the second side

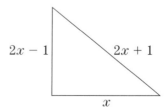

Step 3 Reread and write an equation.

$$c^2 = a^2 + b^2$$
$$(2x + 1)^2 = x^2 + (2x - 1)^2$$

Step 4 Solve the equation.

$$(2x + 1)^2 = x^2 + (2x - 1)^2$$
$$4x^2 + 4x + 1 = x^2 + 4x^2 - 4x + 1$$
$$4x^2 + 4x + 1 = 5x^2 - 4x + 1$$
$$0 = x^2 - 8x$$

$0 = x(x - 8)$	$2x + 1 = 2(8) + 1 = 17$
$x = 0 \text{ or } x - 8 = 0$	$2x - 1 = 2(8) - 1 = 15$
Impossible $x = 8$	

The length of the shortest side is 8 ft.

Step 5 Check.

The sides are 8 ft, 15 ft, and 17 ft. The hypotenuse is 1 ft more than twice 8 ft, and the second side is 1 ft less than twice 8 ft. Also,

$$17^2 = 8^2 + 15^2$$
$$289 = 64 + 225$$
$$289 = 289$$

It checks.

EXAMPLE 6 **The length of one rectangle is three times the width. If the length is increased by 4 feet and the width is decreased by 1 foot, the area will be increased by 24 square feet. Find the dimensions of the original rectangle.**

Solution

Step 1 Read and write givens and unknowns.

We want to find the length and width of the original rectangle. We know that the length of the original rectangle is three times its width. We also know that if we add 4 ft to the length and subtract 1 ft from the width, the new area will be 24 ft^2 more than the original area. (Area = Length · Width.)

Step 2   Begin with "Let x = the unknown."

> Let x = the width of original rectangle
> $3x$ = the length of original rectangle
> $x - 1$ = the width of new rectangle
> $3x + 4$ = the length of new rectangle

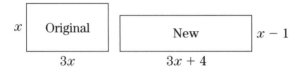

Step 3   Reread and write an equation.

> The new area = 24 more than the original area
> $(3x + 4)(x - 1) = 24 + (x)(3x)$

Step 4   Solve the equation.

$$(3x + 4)(x - 1) = 24 + (x)(3x)$$
$$3x^2 + x - 4 = 24 + 3x^2$$
$$x - 4 = 24$$
$$x = 28 \qquad\qquad 3\,x = 3(28) = 84$$

The dimensions of the original rectangle are 28 ft by 84 ft.
The length is 84 ft and the width is 28 ft.

Step 5   Check.

The area of the original rectangle is $(84)(28)$, or 2352 ft^2.
The dimensions of the new rectangle are 88 ft by 27 ft. Therefore, the area of the new rectangle is $(88)(27)$, or 2376 ft^2. The area of the new rectangle is 24 ft^2 more than the area of the original rectangle. It checks.

PROBLEM SET 3

Solve each of the following geometry problems. Be sure to follow the five-step strategy presented in Chapter 2.

1. The length of a rectangle is four less than twice the width. The perimeter is 46 inches. Find the dimensions of the rectangle.

2. The length of a rectangle is three more than twice the width. The perimeter is 108 centimeters. Find the dimensions of the rectangle.

3. The width of a rectangle is ten less than the length. The perimeter is 52 feet. Find the dimensions of the rectangle.

4. The width of a rectangle is three more than one-half the length. The perimeter is 90 yards. Find the dimensions of the rectangle.

5. The first side of a triangle is one more than three times the third side. The second side is three less than the third side. The perimeter of the triangle is 73 inches. Find the sides of the triangle.

6. The first side of a triangle is two less than four times the second side. The third side is three times the second side. The perimeter of the triangle is 54 feet. Find the sides of the triangle.

7. Lisa is planning a triangular garden for her front yard. She wants the second side to be 2 feet longer than the first side and the third side to be 4 feet longer than the first side. She has 21 feet of fencing to enclose the garden. What will the lengths of each side be?

8. The perimeter of a triangular pool is 43 feet. One side is five more than the shortest side, and the third side is twice the shortest side. Find the lengths of each side.

9. The first angle of a triangle is twice the second angle. The third angle is three times the second angle. Find the measures of the angles of the triangle.

10. In triangle ABC, the measure of angle B is three more than twice the measure of angle A, and the measure of angle C is two more than four times the measure of angle A. Find the measures of the angles of the triangle.

11. In triangle ABC, the measure of angle A is twice the measure of angle B. The measure of angle C is five more than the measure of angle A. Find the measures of the three angles.

12. In triangle ABC, the measure of angle C is one less than three times the measure of angle A. The measure of angle B is four less than the measure of angle A. Find the measures of the three angles.

13. Angle A and angle B are supplementary angles. Angle A is four less than three times angle B. Find the measures of the two angles.

14. Angle A and angle B are supplementary angles. Angle A is five less than four times angle B. Find the measures of the two angles.

15. Angle A and angle B are complementary angles. Angle A is nine more than twice angle B. Find the measures of the two angles.

16. Angle A and angle B are complementary angles. Angle A is two more than three times angle B. Find the measures of the two angles.

17. The perimeter of a square is 27 inches. Find the length of each side.

18. The perimeter of a square is 121 centimeters. Find the length of each side.

19. The circumference of a circle is 25 centimeters. Find the radius to the nearest tenth of a centimeter.

20. The circumference of a circle is 80 millimeters. Find the radius to the nearest tenth of a millimeter.

21. The Canadian Football League requires that rectangular playing fields have specific dimensions. If the perimeter of the field is 350 yards and the length is 20 yards less than twice the width, find the dimensions of the playing fields.

The following problems require algebraic techniques from intermediate algebra to solve.

22. One bicyclist leaves a corner traveling north and rides for 12 miles. A second bicyclist leaves the same corner at the same time and travels east. How far apart are they when the distance between them is three more than twice the distance traveled by the eastbound bicyclist?

23. Sylvia is planning a garden for the corner of her yard. It will be in the shape of a right triangle. She wants the two sides adjacent to the right angle to be 5 feet each. Find the length of the third side to the nearest tenth of a foot.

24. The hypotenuse of a right triangle is 4 centimeters more than three times the shortest side. The other leg is 1 centimeter less than the hypotenuse. Find the length of the hypotenuse.

25. A painter leans a 15-foot ladder against the side of a house. The top of the ladder touches the house at a height that is 3 feet more than the distance from the house to the bottom of the ladder. How far is the bottom of the ladder from the house?

26. The area of a triangle is 45 square inches. The height is three more than twice the base. Find the height of the triangle.

27. The area of a triangle is 33 square inches. The height is one less than twice the base. Find the base of the triangle.

28. Find the radius of a circle to the nearest centimeter if the area is 50 square centimeters.

29. Find the radius of a circle to the nearest inch if the area is 154 square inches.

30. The length of a rectangle is one more than twice its width. Its area is 105 square meters. Find its dimensions.

31. The length of a rectangle is three less than twice its width. Its area is 104 square yards. Find its dimensions.

32. The 1999 Women's World Cup Soccer Tournament was played at the Rose Bowl in Pasadena, California. The perimeter of the rectangular field is 376 yards, and the area of the field is 8352 square yards. Find the dimensions of the field.

33. A baseball diamond is actually a square with sides of 30 yards. Find the distance from home plate to second base (the diagonal of the square). Round your answer to the nearest tenth of a yard.

Coin Problems

Coin problems involve different types of coins, bills, stamps, tickets, or other items that have a value. It is very important in this type of problem to distinguish between *how many* items you have (the *number*) and *how much* the items are worth (their *value*). Sometimes these problems are called number–value problems. We will use a table to organize the information that is given.

In all these problems we will use the principle that *the number times the value of each item equals the total value* of the items. For example, if you have twelve nickels, then the value of these coins is twelve times five cents, or 12($0.05) = $0.60. Remember:

$$\boxed{(\text{Number of items}) \cdot (\text{Value of each item}) = \text{Total value}}$$

We will now use the five-step strategy to solve coin problems.

EXAMPLE 1 **Walter has a total of 26 coins in his pocket. The coins are dimes and nickels and are worth $2.20. How many of each coin does he have?**

Solution

Step 1 Read and write givens and unknowns.

We want to find the number of dimes and the number of nickels. We know that the *total number* of coins is 26 and the *total value* of the coins is $2.20.

Step 2 Begin with "Let x = the unknown."

Let x = the number of nickels
$26 - x$ = the number of dimes

We can organize the givens and unknowns by making a table.

	Number	Value of each (in cents)	Total value
Nickels	x	5	$5(x)$
Dimes	$26 - x$	10	$10(26 - x)$

Step 3   Reread and write an equation.

Total value of the coins = \$2.20   (or 220 cents)

$$5(x) + 10(26 - x) = 220 \quad \text{(This must all be in cents.)}$$

Step 4   Solve the equation.

$$5(x) + 10(26 - x) = 220$$
$$5x + 260 - 10x = 220$$
$$260 - 5x = 220$$
$$-5x = -40$$
$$x = 8 \qquad\qquad 26 - x = 26 - 8 = 18$$

Walter has 8 nickels and 18 dimes.

Step 5   Check.

There are 8 + 18 or 26 coins. The value of the coins is 8(\$0.05) + 18(\$0.10), or \$0.40 + \$1.80 = \$2.20. It checks.

Alternate Solution

Many coin problems can also be solved using a system of linear equations. This first example will be shown this way also.

Step 1   Read and write givens and unknowns.

We want to find the number of dimes and the number of nickels. We know that the *total number* of coins is 26 and the *total value* of the coins is \$2.20.

Step 2   Begin with "Let x = the unknown."

Let x = the number of nickels

y = the number of dimes

	Number	Value of each (in cents)	Total value
Nickels	x	5	$5(x)$
Dimes	y	10	$10(y)$

Step 3 Reread and write an equation.

Total number of coins = 26   Total value of coins = \$2.20 (or 220 cents)

$$x + y = 26 \qquad\qquad 5(x) + 10(y) = 220$$

Step 4 Solve the system of equations.

$$x + \quad y = 26$$
$$5x + 10y = 220$$

Solve for x in the first equation, $y = 26 - x$, and substitute it into the second equation.

$$5(x) + 10(26 - x) = 220$$
$$260 - 5x = 220$$
$$-5x = -40$$
$$x = 8 \qquad y = 26 - x = 26 - 8 = 18$$

Walter has 8 nickels and 18 dimes.

Step 5 Check.

There are $8 + 18$ or 26 coins. The value of the coins is 8(\$0.05) + 18(\$0.10), or \$0.40 + \$1.80 = \$2.20. It checks.

EXAMPLE 2   **Alison went to the post office and bought 72 first-class and postcard stamps for \$21.40. How many of each did she buy if first-class stamps cost 34 cents each and postcard stamps cost 20 cents each?**

Solution

Step 1 Read and write givens and unknowns.

We want to find the number of 34-cent stamps and the number of 20-cent stamps Alison purchased. We know that the total number of stamps was 72 and the total cost was \$21.40.

Step 2 Begin with "Let x = the unknown."

Let x = the number of 34-cent stamps
$72 - x$ = the number of 20-cent stamps

	Number	Value of each (in cents)	Total value
34-cent stamps	x	34	$34(x)$
20-cent stamps	$72 - x$	20	$20(72 - x)$

Step 3 Reread and write an equation.

Total value of the stamps = $21.40   (or 2140 cents)
$$34(x) + 20(72 - x) = 2140$$

Step 4 Solve the equation.

$$34(x) + 20(72 - x) = 2140$$
$$34x + 1440 - 20x = 2140$$
$$14x + 1440 = 2140$$
$$14x = 700$$
$$x = 50 \qquad 72 - x = 72 - 50 = 22$$

Alison purchased 50 first-class stamps and 22 postcard stamps.

Step 5 Check.

The total number of stamps purchased is $50 + 22$ or 72 stamps. The total cost of the stamps is $50(\$0.34) + 22(\$0.22) = \$17.00 + \$4.40 = \$21.40$. It checks.

EXAMPLE 3 **The PTA is selling tickets to a school play. Tickets for adults cost $6.00 each, and children's tickets cost $2.50 each. There are 95 people attending a performance, and the total revenue from ticket sales is $381. How many adults and how many children are attending this performance?**

Solution

Step 1 Read and write givens and unknowns.

We want to find the number of adults and the number of children who are attending the performance. We know that the total number of people is 95 and that the total value of the tickets is $381.

Step 2 Begin with "Let x = the unknown."

Let x = the number of children
$95 - x$ = the number of adults

	Number	Value of each (in dollars)	Total value
Children	x	2.50	$2.50(x)$
Adults	$95 - x$	6	$6(95 - x)$

Step 3   Reread and write an equation.

The total value of the tickets = $381
$$2.50(x) + 6(95 - x) = 381$$

Step 4   Solve the equation.

$$
\begin{aligned}
2.50(x) + 6(95 - x) &= 381 \\
2.50x + 570 - 6x &= 381 \\
570 - 3.50x &= 381 \\
-3.50x &= -189 \\
x &= 54 \qquad 95 - x = 95 - 54 = 41
\end{aligned}
$$

There are 54 children and 41 adults attending the performance.

Step 5   Check.

The total number of people attending the performance is 54 + 41 or 95 people. The total revenue is 54 ($2.50) + 41($6.00) = $135 + 246 = $381. It checks.

EXAMPLE 4   **The cash drawer of Smart Buy Supermarket contains $420 in bills. There are eight more five-dollar bills than ten-dollar bills. The number of one-dollar bills is two more than three times the number of ten-dollar bills. Find the number of each kind of bill in the drawer.**

Solution

Step 1   Read and write givens and unknowns.

We want to find the number of $10, $5, and $1 bills in the cash drawer. We know that the number of $5 bills is eight more than the number of $10 bills. The number of $1 bills is two added to three times the number of $10 bills. We also know that the total value of the bills is $420.

Step 2   Begin with "Let x = the unknown."

Let x = the number of $10 bills
$x + 8$ = the number of $5 bills
$3x + 2$ = the number of $1 bills

	Number	Value of each (in dollars)	Total value
$10 bills	x	10	$10(x)$
$5 bills	$x + 8$	5	$5(x + 8)$
$1 bills	$3x + 2$	1	$1(3x + 2)$

Step 3   Reread and write an equation.

$$\text{The total value of the bills} = \$420$$
$$10(x) + 5(x + 8) + 1(3x + 2) = 420$$

Step 4   Solve the equation.

$$10(x) + 5(x + 8) + 1(3x + 2) = 420$$
$$10x + 5x + 40 + 3x + 2 = 420$$
$$18x + 42 = 420$$
$$18x = 378 \qquad x + 8 = 21 + 8 = 29$$
$$x = 21 \qquad 3x + 2 = 3(21) + 2 = 65$$

The cash drawer contains 21 $10 bills, 29 $5 bills, and 65 $1 bills.

Step 5   Check.

The number of $5 bills is eight more than the number of $10 bills. The number of $1 bills is two more than three times the number of $10 bills. The total value of the bills is $21(\$10) + 29(\$5) + 65(\$1) = \$210 + \$145 + \$65 = \$420$. It checks.

PROBLEM SET 4

Solve each of the following coin problems. Be sure to follow the five-step strategy. Make a table for each problem.

1. Nancy has 26 coins in her pocket. The coins are dimes and quarters and are worth $3.65. How many dimes and quarters does she have?

2. Barry has 22 coins in his pocket. The coins are quarters and dimes and are worth $4.75. How many quarters and dimes does he have?

3. A bank teller has 24 coins in his cash drawer. The coins are quarters and nickels and are worth $3. How many of each coin does he have?

4. A bank teller has 27 coins in her cash drawer. The coins are quarters and nickels and are worth $2.75. How many of each coin does she have?

5. Juan has $4.80 in quarters and dimes. If he has a total of 27 coins, how many dimes does he have?

6. Samantha finds a change purse at the market. The purse contains 19 coins worth $3.40. If the purse contains only dimes and quarters, how many dimes are in the purse?

7. Jose went to the post office for first-class and postcard stamps. He bought 22 stamps and paid $6.50 for them. If first-class stamps cost 34 cents each and postcard stamps cost 20 cents each, how many of each did he buy?

8. Terry went to the post office for first-class and postcard stamps. He bought 60 stamps and paid $17.88 for them. If first-class stamps cost 34 cents each and postcard stamps cost 20 cents each, how many of each did he buy?

9. Marilyn went to the post office to buy two kinds of stamps. She bought 20 stamps and paid $6 for them. If one type of stamp costs 45 cents each and the other costs 20 cents each, how many of each did she get?

10. Jennifer went to the post office to buy two kinds of stamps. She bought 22 stamps and paid $7.90 for them. If one type of stamp costs 45 cents each and the other costs 20 cents each, how many of each did she get?

11. The Key Club is selling tickets to a fund-raising concert. Tickets for adults cost $18 each, and tickets for children cost $14 each. There are 93 people at the concert and the total revenue from ticket sales is $1574. How many adults and how many children are attending the concert?

12. The Los Osos Garden Club held its annual garden show. Tickets for adults cost $8 each, and tickets for children cost $5 each. There are 520 people who attended the show, and the total revenue collected was $3245. How many adults and how many children attended the show?

13. There were 115 people who attended the Eagles basketball game. Tickets for students sold for $4 each, and tickets for nonstudents sold for $7 each. If the total ticket sales were $553, how many students attended the game?

14. A promoter sold 250 tickets to a concert. An adult's ticket cost $25, and a child's ticket cost $15. If the total ticket sales were $5620, how many of each type of ticket were sold?

15. Sky Airways sells one-way tickets from Los Angeles to Portland to adults for $100 and to children for $75 each. The total collected on a particular flight that carried 60 people was $5825. How many adults were on this flight?

16. Admission to Showtime Cinema costs $7 for adults and $4 for children. If 78 people attended a movie and the cashier collected $468, how many adults were at the show?

17. Jim is saving for a ski trip. He has saved $139 in ten-dollar, five-dollar, and one-dollar bills. The number of five-dollar bills is two more than three times

the number of ten-dollar bills. The number of one-dollar bills is one less than the number of ten-dollar bills. Find the number of each type of bill.

18. Sharon has $106 in her wallet. She has twice as many ten-dollar bills as five-dollar bills and two more one-dollar bills than five-dollar bills. Find the number of each type of bill in her wallet.

19. Marvin has $3.86 in quarters, dimes, and pennies. The number of pennies is twice the number of dimes, and there are 23 coins in all. How many dimes does he have?

20. A cash register contains $362. It contains the same number of one-dollar bills as five-dollar bills. Also, there are twice as many ten-dollar bills as five-dollar bills and two more twenty-dollar bills than five-dollar bills. How many five-dollar bills are there?

21. A bank teller has 32 coins in his cash drawer. There are three times as many quarters as nickels. The number of dimes is two more than twice the number of nickels. If the coins are worth $5.20, how many of each coin does he have?

22. The Bay Theater charges $8 for adults and $5 for children. At the noon performance, there were twice as many children as adults and the total collected was $576. How many adults and how many children attended the performance?

23. The Sunday, March 30, 2003, matinee performance of "The Producers" at the Orpheum Theatre in San Francisco was sold out. The capacity of the orchestra is 1289, of the mezzanine is 622, and of the balcony is 292. The price of a mezzanine ticket is $20 more than a balcony ticket. The price of an orchestra ticket is $11 less than twice the price of a balcony ticket. If the total receipts from ticket sales for the performance was $158,893, find the price of a balcony ticket.

Solve the following problems using a system of linear equations.

24. Jeremy has 19 coins in his pocket. The coins are dimes and quarters and are worth $3.55. How many of each does he have?

25. A bank teller has 42 coins in her cash drawer. The coins are nickels, dimes, and quarters. There is an equal number of nickels and dimes, and the coins are worth $4.55. Find the number of each.

26. The Friends of the Library are having a used-book sale. Paperback books cost $1.50 each, and hardcover books cost $2.50 each. Jack bought 14 books and spent $26. How many of each did he buy?

27. The Royal Oak Motel rents rooms for the night for $60 for a single and $80 for a double. On Thursday night, they rented 31 rooms and collected $2260. How many singles and how many doubles were rented that night?

28. Sharon has quarters and dimes in her purse. She accidentally spills the 79 coins on the floor. While picking up the money, she discovers that she has $14.20. How many quarters and how many dimes does she have?

29. The Alpine Ski Club is selling tickets to the annual show. Tickets for students are $8 each and for nonstudents $15 each. There are 112 people who attend the show, and the total ticket sales are $1099. How many of each was sold?

30. The Black and Gold Booster Club is holding a chicken barbecue. Tickets for adults are $8 each, and tickets for children are $5 each. The event brought in $578 for the 85 tickets sold. How many of each ticket was sold?

31. Joshua went to the post office to buy first-class and postcard stamps. He bought 58 stamps and paid $18.60 for them. If first-class stamps cost 34 cents each and postcard stamps cost 20 cents each, how many of each did he buy?

Mixture Problems

Mixture problems involve mixing different substances together. We will concentrate on two types of mixture problems. The first involves the percents of solutions, and the second involves the prices of quantities being mixed together.

In percent-mixture problems, we use the principle that the amount times the percent of the quantity equals the total amount of that quantity. For example, if we have 8 quarts of a 25% sulfuric acid solution, then the amount (8) times the percent of sulfuric acid (25%) equals the total amount of sulfuric acid. We have $8(0.25) = 2$, so there are 2 quarts of sulfuric acid in this solution.

Remember:

$$\boxed{\text{Amount} \cdot \text{Percent} = \text{Total amount}}$$

If we mix together two sulfuric acid solutions, then the amount of sulfuric acid in the first added to the amount of sulfuric acid in the second equals the amount of sulfuric acid in the mixture. Charts or diagrams are very helpful in solving mixture problems. In the following four examples, we will use a mixture diagram to organize the information as an equation:

$$\boxed{\text{First solution}} + \boxed{\text{Second solution}} = \boxed{\text{Mixture}}$$

EXAMPLE 1   A chemist has a solution that is 75% sulfuric acid and a solution that is 25% sulfuric acid. How much of each

should she use to obtain 40 milliliters of a solution that is 45% sulfuric acid?

Solution

Step 1   Read and write givens and unknowns.

We want to find the amount of 75% sulfuric acid solution and the amount of 25% sulfuric acid solution to be mixed together. We know that she wants 40 ml of the mixture that will be 45% sulfuric acid.

Step 2   Begin with "Let x = the unknown."

Let x = the amount of 75% sulfuric acid solution
$40 - x$ = the amount of 25% sulfuric acid solution

We can organize the givens and unknowns by making a mixture diagram:

1st solution		2nd solution		Mixture
x ml of 75% solution	$+$	$(40 - x)$ ml of 25% solution	$=$	40 ml of 45% solution

Step 3   Reread and write an equation.

Amount of sulfuric acid in 1st solution		Amount of sulfuric acid in 2nd solution		Amount of sulfuric acid in mixture
$0.75x$	$+$	$0.25(40 - x)$	$=$	$0.45(40)$

Step 4   Solve the equation.

$$0.75x + 0.25(40 - x) = 0.45(40)$$
$$0.75 + 10 - 0.25x = 18$$
$$0.50x = 8$$
$$x = 16 \qquad 40 - x = 40 - 16 = 24$$

The chemist should mix 16 ml of 75% solution with 24 ml of 25% solution to obtain 40 ml of 45% solution.

Step 5   Check.

16 ml of 75% solution contains $(0.75)(16)$ or 12 ml of sulfuric acid.
24 ml of 25% solution contains $(0.25)(24)$ or 6 ml of sulfuric acid.
40 ml of 45% solution contains $(0.45)(40)$ or 18 ml of sulfuric acid.
Since $16 + 24 = 40$ and $12 + 6 = 18$, it checks.

Alternate Solution

Many mixture problems can also be solved using a system of linear equations. This first example will be shown this way also.

Step 1   Read and write givens and unknowns.

We want to find the amount of 75% sulfuric acid solution and the amount of 25% sulfuric acid solution to be mixed together. We know that she wants 40 ml of the mixture that will be 45% sulfuric acid.

Step 2   Begin with "Let x = the unknown."

Let x = the amount of 75% sulfuric acid solution
    y = the amount of 25% sulfuric acid solution

1st solution		2nd solution		Mixture
x ml of 75% solution	+	y ml of 25% solution	=	40 ml of 45% solution

Step 3   Reread and write an equation.

Amount of sulfuric acid in 1st solution + Amount of sulfuric acid in 2nd solution = Amount of sulfuric acid in in mixture

$$0.75x \quad + \quad 0.25y \quad = \quad 0.45(40)$$

Also, the total amount of the mixture = 40 milliliters
$$x + y = 40$$

Step 4   Solve the system of equations.

$$x + y = 40$$
$$0.75x + 0.25y = 0.45(40)$$

Solve for y in the first equation, $y = 40 - x$, and substitute it into the second equation.

$$0.75x + 0.25(40 - x) = 0.45\,(40)$$
$$0.75x + 10 - 0.25x = 18$$

$$0.50x + 10 = 18 \qquad\qquad x + y = 40$$
$$0.50x = 8 \qquad\qquad 16 + y = 40$$
$$x = 16 \qquad\qquad y = 24$$

The chemist should mix 16 ml of 75% solution with 24 ml of 25% solution to obtain 40 ml of 45% solution.

Step 5 Check.

16 ml of 75% solution contains $(0.75)(16)$ or 12 ml of sulfuric acid.
24 ml of 25% solution contains $(0.25)(24)$ or 6 ml of sulfuric acid.
40 ml of 45% solution contains $(0.45)(40)$ or 18 ml of sulfuric acid.
Since $16 + 24 = 40$ and $12 + 6 = 18$, it checks.

EXAMPLE 2 **How many liters of a 40% alcohol solution should be added to 40 liters of a 10% alcohol solution to obtain a 20% alcohol solution?**

Solution

Step 1 Read and write givens and unknowns.

We want to find the amount of 40% alcohol solution. We know that there are 40 liters of the 10% alcohol solution and that the mixture will be a 20% alcohol solution.

Step 2 Begin with "Let x = the unknown."

Let x = the amount of 40% alcohol solution

1st solution		2nd solution		Mixture
x liters of 40% solution	$+$	40 liters of 10% solution	$=$	$(40 + x)$ liters of 20% solution

Note: Because x liters of the first solution are added to 40 liters of the second solution, we obtain $(40 + x)$ liters of the mixture.

Step 3 Reread and write an equation.

Amount of alcohol in 1st solution		Amount of alcohol in 2nd solution		Amount of alcohol in mixture
$0.40x$	$+$	$0.10(40)$	$=$	$0.20(40 + x)$

Step 4   Solve the equation.

$$0.40x + 0.10(40) = 0.20(40 + x)$$
$$0.40x + 4 = 8 + 0.20x$$
$$0.20x = 4$$
$$x = 20 \qquad\qquad 40 + x = 40 + 20 = 60$$

Twenty liters of 40% alcohol solution should be mixed with 40 liters of 10% solution to obtain 60 liters of 20% alcohol solution.

Step 5   Check.

20 liters of 40% solution contains $(0.40)(20)$ or 8 liters of alcohol.
40 liters of 10% solution contains $(0.10)(40)$ or 4 liters of alcohol.
60 liters of 20% solution contains $(0.20)(60)$ or 12 liters of alcohol.
Since $20 + 40 = 60$ and $8 + 4 = 12$, it checks.

EXAMPLE 3   **Christi wants to dilute a 10% soap solution with water to make 10 gallons of a 3% soap solution. How much water and how much 10% solution should she mix together?**

Solution

Step 1   Read and write givens and unknowns.

We want to find the amount of water and the amount of 10% solution to be mixed together. We know that she wants to mix a 10% soap solution with water to make 10 gallons of a 3% soap solution. We also know that water is a 0% soap solution. In other words, there is no soap in plain water.

Step 2   Begin with "Let $x =$ the unknown."

Let $x =$ the amount of water (or 0% soap solution)
$10 - x =$ the amount of 10% soap solution

1st solution		2nd solution		Mixture
x gal of water or 0% solution	$+$	$(10 - x)$ gal of 10% solution	$=$	10 gal of 3% solution

Step 3   Reread and write an equation.

| Amount of soap in 1st solution | + | Amount of soap in 2nd solution | = | Amount of soap in mixture |

$$0(x) + 0.10(10 - x) = 0.03(10)$$

Step 4   Solve the equation.

$$0(x) + 0.10(10 - x) = 0.03(10)$$
$$0 + 1 - 0.10x = 0.3$$
$$-0.10x = -0.7$$
$$x = 7 \qquad 10 - x = 10 - 7 = 3$$

Christi should add 7 gallons of water to 3 gallons of a 10% soap solution to obtain 10 gallons of a 3% soap solution.

Step 5   Check.

7 gallons of water has no soap. 3 gallons of 10% soap solution has 0.10(3) or 0.3 gallon of soap. 10 gallons of 3% soap solution has 0.03(10) or 0.3 gallon of soap. Since $7 + 3 = 10$ and $0 + 0.3 = 0.3$, it checks.

EXAMPLE 4   **A nurse needs a 50% boric acid solution for his patient. How much pure boric acid should he add to 8 ounces of a 10% boric acid solution to obtain a 50% boric acid solution?**

Solution

Step 1   Read and write givens and unknowns.

We want to find the amount of pure boric acid. We know that the nurse has 8 ounces of a 10% solution that will be added to the *pure* boric acid to make a 50% solution. We also know that pure boric acid is a 100% solution.

Step 2   Begin with "Let x = the unknown."

Let x = the amount of pure boric acid (100% solution)

1st solution		2nd solution		Mixture
x oz of pure boric acid or 100% solution	+	8 oz of 10% solution	=	$(x + 8)$ oz of 50% solution

Step 3 Reread and write an equation.

Amount of boric acid in 1st solution	+	Amount of boric acid in 2nd solution	=	Amount of boric acid in mixture
$1.00x$	$+$	$0.10(8)$	$=$	$0.50(x + 8)$

Step 4 Solve the equation.

$$1.00x + 0.10(8) = 0.50(x + 8)$$
$$1.0x + 0.8 = 0.5x + 4$$
$$0.5x = 3.2$$
$$x = 6.4 \qquad\qquad x + 8 = 6.4 + 8 = 14.4$$

The nurse should add 6.4 ounces of pure boric acid to the 8 ounces of 10% solution to obtain 14.4 ounces of the 50% boric acid solution.

Step 5 Check.

6.4 ounces of pure boric acid has 6.4 ounces of boric acid.
8 ounces of a 10% solution has $0.10(8)$ or 0.8 ounce of boric acid.
14.4 ounces of a 50% solution has $0.50(14.4)$ or 7.2 ounces of boric acid.
Since $6.4 + 8 = 14.4$ and $6.4 + 0.8 = 7.2$, it checks.

The second type of mixture problem is similar to coin problems. We use the principle that the number multiplied by the price (or value) of each equals the total price (or total value). In these problems, the value of the total mixture must be equal to the sum of the values of the two quantities being mixed together.

Remember from coin problems:

$$\boxed{\text{Number} \cdot \text{Value of each} = \text{Total value}}$$

EXAMPLE 5 **Cara has 15 pounds of cashews that sell for \$5.25 per pound. If peanuts sell for \$2.50 per pound, how many pounds of peanuts should she add to the cashews to obtain a mixture that will sell for \$3.75 per pound?**

Solution

Step 1 Read and write givens and unknowns.

We want to find the number of pounds of peanuts. We know that Cara has 15 pounds of cashews that sell for $5.25 per pound and that the mixture of the cashews and peanuts will sell for $3.75 per pound.

Step 2 Begin with "Let x = the unknown."

Let x = the number of pounds of peanuts

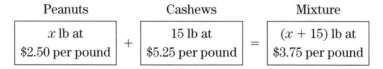

Peanuts		Cashews		Mixture
x lb at $2.50 per pound	$+$	15 lb at $5.25 per pound	$=$	$(x + 15)$ lb at $3.75 per pound

Step 3 Reread and write an equation.

$$\text{Value of peanuts} + \text{Value of cashews} = \text{Value of mixture}$$

$$2.50x + 5.25(15) = 3.75(x + 15)$$

Step 4 Solve the equation.

$$2.50x + 5.25(15) = 3.75(x + 15)$$
$$2.50x + 78.75 = 3.75x + 56.25$$
$$22.50 = 1.25x$$
$$18 = x \qquad\qquad x + 15 = 18 + 15 = 33$$

Cara should add 18 pounds of peanuts to the 15 pounds of cashews to obtain 33 pounds of the mixture.

Step 5 Check.

18 pounds of peanuts is worth 18($2.50) or $45.
15 pounds of cashews is worth 15($5.25) or $78.75.
33 pounds of the mixture is worth 33($3.75) or $123.75.
Since 18 + 15 = 33 and $45 + $78.75 = $123.75, it checks.

EXAMPLE 6 **Stephanie wishes to mix café mocha worth $5 per pound with regular coffee worth $3 per pound to make 8 pounds of a mixture to sell for $3.50 per pound. How many pounds of each should she use?**

Solution

Step 1 Read and write givens and unknowns.

We want to find the number of pounds of café mocha priced at $5 per pound and regular coffee priced at $3 per pound. We know that Stephanie wants to make 8 pounds of a mixture to sell for $3.50 per pound.

Step 2 Begin with "Let x = the unknown."

Let x = the number of pounds of café mocha
$8 - x$ = the number of pounds of regular coffee

Café mocha		Regular coffee		Mixture
x lb at $5.00 per pound	$+$	$(8 - x)$ lb at $3.00 per pound	$=$	8 lb at $3.50 per pound

Step 3 Reread and write an equation.

$$\begin{array}{ccc} \text{Value of} & \text{Value of} & \text{Value of} \\ \text{café mocha} & + \quad \text{regular coffee} & = \quad \text{mixture} \end{array}$$

$$5.00x \quad + \quad 3.00(8 - x) \quad = \quad 3.50(8)$$

Step 4 Solve the equation.

$$5.00x + 3.00(8 - x) = 3.50(8)$$
$$5x + 24 - 3x = 28$$
$$2x + 24 = 28$$
$$2x = 4$$
$$x = 2 \qquad\qquad 8 - x = 8 - 2 = 6$$

Stephanie should use 2 pounds of café mocha and 6 pounds of regular coffee to make 8 pounds of the mixture.

Step 5 Check.

2 pounds of café mocha is worth 2($5.00) or $10.
6 pounds of regular coffee is worth 6($3.00) or $18.
8 pounds of the mixture is worth 8($3.50) or $28.
Since $2 + 6 = 8$ and $10 + 18 = 28$, it checks.

The next example is more difficult and would be found in an intermediate algebra course. It involves taking away some fluid and replacing it.

EXAMPLE 7 **A radiator contains 8 quarts of a 20% antifreeze solution. How much should be drained and replaced with pure**

antifreeze to produce a 30% antifreeze solution in the radiator?

Solution

Step 1 Read and write givens and unknowns.

We want to find the number of quarts of a 20% antifreeze solution that should be drained and replaced with pure antifreeze. We know that we begin with 8 quarts of a 20% solution and the result will be 8 quarts of a 30% solution. We also know that pure antifreeze is a 100% solution.

Step 2 Begin with "Let x = the unknown."

Let x = the number of quarts of 20% antifreeze solution to be drained
x = the number of quarts of pure antifreeze to be added

Original solution	Drained solution	Replaced solution	Final solution
8 qt of 20% solution	x qt of 20% solution	x qt of 100% antifreeze	8 qt of 30% solution

$$\text{Original} - \text{Drained} + \text{Replaced} = \text{Final}$$

Step 3 Reread and write an equation.

Amount of antifreeze in radiator	$-$	Amount of antifreeze drained	$+$	Amount of antifreeze added	$=$	Final amount of antifreeze
$0.20(8)$	$-$	$0.20x$	$+$	$1.00x$	$=$	$0.30(8)$

Step 4 Solve the equation.

$$0.20(8) - 0.20x + 1.00x = 0.30(8)$$
$$1.6 - 0.2x + 1.0x = 2.4$$
$$0.8x + 1.6 = 2.4$$
$$0.8x = 0.8$$
$$x = 1$$

One quart of a 20% antifreeze solution should be drained from the radiator and replaced with pure antifreeze to produce a 30% antifreeze solution in the radiator.

Step 5 Check.

Amount of antifreeze in radiator is 0.20(8) or 1.6 quarts.
Amount of antifreeze drained is 0.20(1) or 0.20 quart.

Amount of antifreeze added is 1.00(1) or 1 quart.

Final amount of antifreeze is 0.30(8) or 2.4 quarts.

Since $1.6 - 0.20 + 1 = 2.4$, it checks.

PROBLEM SET 5

Solve each of the following mixture problems. Be sure to follow the five-step strategy. Make a mixture diagram for each problem.

1. How many liters of a 10% salt solution should be added to 80 liters of a 35% salt solution to obtain a mixture that is a 30% salt solution?

2. How many pounds of cheese with 45% fat content should be mixed with 6 pounds of cheese with 20% fat content to obtain cheese with 30% fat content?

3. A farmer has a 10% insecticide solution and a 50% insecticide solution. How much of each should he mix to obtain 40 gallons of a 35% insecticide solution?

4. How many ounces of a 4% alcohol solution and a 1% alcohol solution should be mixed together to obtain 39 ounces of a 2% alcohol solution?

5. A photographer has 12 quarts of 5% acetic acid solution. How much 10% acetic acid solution should she add to obtain a mixture that is 7% acetic acid?

6. How many pounds of cottage cheese that is 4% butterfat should be mixed with 40 pounds of cottage cheese that is 1% butterfat to make low-fat cottage cheese that is 2% butterfat?

7. A chemist has a 10% acid solution and a 50% acid solution. How many ounces of each should she mix to obtain 24 ounces of a 35% acid solution?

8. Carl has one fruit drink that is 5% juice and another that is 10% juice. How many ounces of each should he mix to obtain 300 ounces of fruit drink that is 7% juice?

9. Ken wants to dilute a 35% boric acid solution with water to make 28 ounces of a 10% boric acid solution. How much water and how much 35% boric acid solution should he mix together?

10. Erika wants to dilute a 12% salt solution with water to make 36 ounces of an 8% salt solution. How much water and how much 12% salt solution should she mix together?

11. How much water should be added to 4 quarts of a 10% soap solution to obtain a 4% soap solution?

12. How much water should be added to 32 cc of a 25% boric acid solution to obtain a 10% boric acid solution?

13. How much water should be added to 60 ounces of a 20% salt solution to obtain a 15% salt solution?

14. How much water should be added to 10 gallons of a 30% soap solution to obtain a 12% soap solution?

15. How much pure alcohol should be added to 150 cc of a 15% alcohol solution to obtain a 25% alcohol solution?

16. How much pure boric acid should be added to 88 cc of an 8% boric acid solution to obtain a 12% boric acid solution?

17. A scientist needs a 30% sodium-iodine solution for his research. He has a 10% sodium-iodine solution and pure sodium-iodine. How much of each should he mix together to obtain 108 milliliters of the 30% sodium-iodine solution?

18. A farmer needs a 35% disinfectant solution to spray her crops. She has a 10% disinfectant solution and pure disinfectant. How much of each should she mix together to obtain 36 quarts of the 35% disinfectant solution?

19. How much pure hydrogen peroxide should be added to 16 ounces of a 25% hydrogen peroxide solution to obtain a 40% hydrogen peroxide solution?

20. A radiator contains 8 quarts of a 20% antifreeze solution. How much should be drained and replaced with pure antifreeze to produce a 45% antifreeze solution in the radiator? (See Example 7.)

21. Allison has jelly beans that sell for $0.75 per pound and small chocolate eggs that sell for $2.25 per pound. How many pounds of each should she use to obtain 12 pounds of an Easter mix to sell for $1.75 per pound?

22. Jim has chocolate-covered raisins that sell for $2.50 per pound and chocolate-covered nuts that sell for $3.25 per pound. How many pounds of each should he use to obtain 27 pounds of a bridge mix to sell for $3 per pound?

23. A garden shop mixes Kentucky bluegrass seed that sells for $6.50 per pound with rye grass seed that sells for $3 per pound. How much of each should be combined to obtain 14 pounds of a mixture that sells for $5 per pound?

24. How much fuel that sells for $0.85 per gallon should be added to 13 gallons of fuel that sells for $1.75 per gallon to obtain a mixture that sells for $1.50 per gallon?

25. How many gallons of paint that sells for $3 per gallon should be added to 11 gallons of paint that sells for $19 per gallon to obtain a mixture that sells for $14 per gallon?

Use a system of equations to solve the following problems.

26. The Sweet Shoppe makes trail mix with mixed nuts that sell for $3.50 per pound and M & M's that sell for $5 per pound. How many pounds of each should be combined to make 15 pounds of trail mix that sells for $4 per pound?

27. A medical researcher needs a 30% sodium-iodine solution. How many ounces of a 10% sodium-iodine solution and a 60% sodium-iodine solution should he mix to obtain 15 ounces of a 30% solution?

28. A nurse has a 25% hydrogen peroxide solution and a solution that is 60% hydrogen peroxide. How much of each should he use to obtain 20 ounces of a 46% hydrogen peroxide solution?

29. How much water must be added to 3 gallons of a 4% insecticide solution to reduce the concentration of insecticide to 3%?

30. Erika wishes to mix peanuts worth $3.25 per pound with raisins that sell for $2.50 per pound to make 9 pounds of a mixture to sell for $3 per pound. How many pounds of each should she use?

6

Finance Problems

Several types of finance problems are included in this chapter. In general, finance problems involve money.

Simple Interest Problems

The first type of finance problem we will solve is simple interest problems. This type deals with a loan or an investment of money in stocks, bonds, or bank accounts. To calculate the amount of money earned (*interest*), we multiply the amount of money invested (*principal*) by the rate of interest (*rate*) and the amount of time (*time*) it is left in the account.

$$\boxed{\text{Interest} = \text{Principal} \cdot \text{Rate} \cdot \text{Time}}$$

Things to remember when solving simple interest problems

1. In most of these problems, the time is 1 year. Often the time is omitted because of the multiplicative identity property of one. However, remember to include the time if it is anything other than 1 year.
2. The rate of interest and the time must agree. In other words, if the rate is per month, then the time must be in months. If the rate is per year, the time must be in years.
3. The rate is usually given as a percent and must be changed to a decimal.
4. In investments, we use a positive interest rate to represent a gain and a negative interest rate to represent a loss.
5. In these problems, the amount of money earned, the income

earned, the interest received, and the return on your investment all refer to the same thing.

EXAMPLE 1   **Jose inherited $30,000 from his aunt. He invested part of it in stocks and the rest in bonds. The stocks paid an annual interest rate of 5%, and the bonds paid 4% annually. If his total annual income from these two investments was $1315, how much did he invest in stocks and how much in bonds?**

Solution

Step 1   Read and write givens and unknowns.

We want to find the amount Jose invested in stocks (at 5%) and the amount he invested in bonds (at 4%). We know that the total amount of money invested (the total principal) is $30,000 and the total annual income (the total interest) is $1315. Because we are looking at annual income, the time is 1 year and we can omit it.

Step 2   Begin with "Let x = the unknown."

Let x = the amount invested in stocks
$30,000 - x$ = the amount invested in bonds

We can organize the givens and unknowns by making a table.

	Principal	Rate	Interest
Stocks	x	0.05	$0.05x$
Bonds	$30,000 - x$	0.04	$0.04(30,000 - x)$

Step 3   Reread and write an equation.

The total interest = $1315
$$0.05x + 0.04(30,000 - x) = 1315$$

Step 4   Solve the equation.

$$0.05x + 0.04(30,000 - x) = 1315$$
$$0.05x + 1200 - 0.04x = 1315$$
$$0.01x + 1200 = 1315$$
$$0.01x = 115 \qquad 30,000 - x = 30,000 - 11,500$$
$$x = 11,500 \qquad\qquad\quad = 18,500$$

Jose invested $11,500 in stocks and $18,500 in bonds.

Step 5   Check.

The total invested is $11,500 + $18,500 or $30,000. The interest on the stocks is 0.05($11,500) or $575. The interest on the bonds is 0.04($18,500) or $740. The total annual interest is $575 + $740 or $1315. It checks.

Alternate Solution

This example can also be solved using a system of linear equations.

Step 1   Read and write givens and unknowns.

We want to find the amount Jose invested in stocks (at 5%) and the amount he invested in bonds (at 4%). We know that the total amount of money invested (the total principal) is $30,000 and the total annual income (the total interest) is $1315. Because the time is 1 year, we can omit it.

Step 2   Begin with "Let x = the unknown."

Let x = the amount invested in stocks
y = the amount invested in bonds

	Principal	Rate	Interest
Stocks	x	0.05	$0.05x$
Bonds	y	0.04	$0.04y$
Total	30,000		1315

Step 3   Reread and write an equation.

The total principal = $30,000   The total interest = $1315
$$x + y = 30{,}000 \qquad 0.05x + 0.04y = 1315$$

Step 4   Solve the system of equations.

$$x + \quad y = 30{,}000$$
$$0.05x + 0.04y = 1315$$

Solve for y in the first equation, $y = 30{,}000 - x$, and substitute it into the second equation.

$$0.05x + 0.04(30{,}000 - x) = 1315$$
$$0.05x + 1200 - 0.04x = 1315$$
$$0.01x + 1200 = 1315$$
$$0.01x = 115 \qquad y = 30{,}000 - x = 30{,}000 - 11{,}500$$
$$x = 11{,}500 \qquad = 18{,}500$$

Jose invested $11,500 in stocks and $18,500 in bonds.

Step 5 Check.

The total invested is $11,500 + $18,500 or $30,000. The interest on the stocks is 0.05($11,500) or $575. The interest on the bonds is 0.04($18,500) or $740. The total annual interest is $575 + $740 or $1315. It checks.

EXAMPLE 2 **Mr. Franklin wants to invest a sum of money in an investment so that the annual interest pays for a yearly scholarship in memory of his wife. The investment pays 5% simple interest, and the annual scholarship will be $2500. How much does Mr. Franklin have to invest?**

Solution

Step 1 Read and write givens and unknowns.

We want to find the amount of money Mr. Franklin has to invest (the principal). We know that the rate of interest is 5% and the annual interest is $2500.

Step 2 Begin with "Let x = the unknown."

Let x = the amount invested at 5%

Principal	Rate	Interest
x	0.05	$0.05x$

Step 3 Reread and write an equation.

The annual interest = $2500
$$0.05x = 2500$$

Step 4 Solve the equation.

$$0.05x = 2500$$
$$x = 50,000$$

Mr. Franklin has to invest $50,000.

Step 5 Check.

The annual interest is 0.05($50,000) or $2500. It checks.

EXAMPLE 3 **How long will it take an investment of $100 to triple if it is invested at 4% simple interest?**

Solution

Step 1 Read and write givens and unknowns.

We want to find the time it takes the investment to triple. We know that the principal is \$100 and the rate is 4%. If the money will triple, then the total interest is \$200 because we have to end up with \$300.

Step 2 Begin with "Let x = the unknown."

Let x = the number of years it will take for the money to triple

Principal	Rate	Time	Interest
100	0.04	x	$100(0.04)(x)$

Step 3 Reread and write an equation.

The total interest = \$200
$$100(0.04)(x) = 200$$

Step 4 Solve the equation.

$$100(0.04)(x) = 200$$
$$4x = 200$$
$$x = 50$$

It will take 50 years for \$100 to triple.

Step 5 Check.

If we invest \$100 for 50 years at 4%, we will make $100(0.04)(50) = \$200$ in interest. We will then have the original principal, \$100, plus the interest, \$200, or \$300, which is triple the amount we started with. It checks.

EXAMPLE 4 **Sherry has borrowed some money from her mother at 3% simple interest and some money from her aunt at 4% simple interest. She borrowed twice as much from her mother as she did from her aunt. At the end of 1 year, the total annual interest due is \$47.50. How much did Sherry borrow from each person?**

Solution

Step 1 Read and write givens and unknowns.

We want to find the amounts Sherry borrowed from her mother at 3% and from her aunt at 4%. We know that the amount borrowed from her mother is twice the amount borrowed from her aunt. We know that the total annual interest from both loans is \$47.50.

Step 2 Begin with "Let x = the unknown."

Let x = the amount borrowed from her aunt at 4%
 $2x$ = the amount borrowed from her mother at 3%

	Principal	Rate	Interest
Aunt	x	0.04	$0.04x$
Mother	$2x$	0.03	$0.03(2x) = 0.06x$

Step 3 Reread and write an equation.

The total annual interest = $47.50
$$0.04x + 0.06x = 47.50$$

Step 4 Solve the equation.

$$0.04x + 0.06x = 47.50$$
$$0.10x = 47.50$$
$$x = 475 \qquad\qquad 2x = 2(475) = 950$$

Sherry borrowed $475 from her aunt and $950 from her mother.

Step 5 Check.

The interest due to her aunt is $475(0.04) or $19. The interest due to her mother is $950(0.03) or $28.50. The total interest due is $19 + $28.50, or $47.50. It checks.

The next example involves a gain and a loss and might be found in an intermediate algebra course.

EXAMPLE 5 **Amy received $10,000 as a graduation gift from her grandparents. She invested it in stocks and bonds. This year she received a 4% annual return on the bonds, but the stocks lost 2% of her investment. If her total annual return from these two investments was $265, how much did she invest in each?**

Solution

Step 1 Read and write givens and unknowns.

We want to find the amount invested in bonds and the amount invested in stocks. We know that the rate for the bonds was 4% and the rate for stocks was -2% (because she lost money on the stocks). We also know that her total principal was $10,000 and her total annual interest was $265.

Step 2 Begin with "Let x = the unknown."

Let x = the amount invested in bonds

$10,000 - x$ = the amount invested in stocks

	Principal	**Rate**	**Interest**
Bonds	x	0.04	$0.04x$
Stocks	$10,000 - x$	-0.02	$-0.02(10,000 - x)$

Step 3 Reread and write an equation.

The total annual interest = \$265

$$0.04x + (-0.02)(10,000 - x) = 265$$

Step 4 Solve the equation.

$$
\begin{aligned}
0.04x + (-0.02)(10,000 - x) &= 265 \\
0.04x - 200 + 0.02x &= 265 \\
0.06x - 200 &= 265 \\
0.06x &= 465 \\
x &= 7750 \qquad 10,000 - x = 10,000 - 7750 = 2250
\end{aligned}
$$

Amy invested \$7750 in bonds and \$2250 in stocks.

Step 5 Check.

The return on the bonds was \$7750 (0.04), or \$310. The return on the stocks was \$2250 (-0.02), or $-$\$45. The total return was \$310 $-$ \$45, or \$265. It checks.

Break-Even Analysis

In a business, when the money taken in (the *revenue*) equals the money going out (the *cost*), the company will break even. This is called the **break-even point**:

$$\boxed{\text{Revenue} = \text{Cost}}$$

To compute the costs in a manufacturing business, we use the formula $C = f + vx$, where C is the total cost, f is the fixed cost, v is the cost per item produced, and x is the number of items produced. To compute the revenue, we use the formula $R = px$, where R is the revenue, p is the price per item, and x is the number of items sold.

EXAMPLE 6   **Sun Electronics produces calculators. It costs \$12 in parts and labor to produce each calculator. The fixed costs are \$1155 per month. If Sun sells each calculator for \$45, how many calculators do they have to produce and sell each month to break even?**

Solution

Step 1   Read and write givens and unknowns.

We want to find the number of calculators that have to be produced and sold each month to break even. We know that the fixed cost is \$1155, the cost to produce each calculator is \$12, and the price per calculator is \$45.

Step 2   Begin with "Let x = the unknown."

Let x = the number of calculators produced and sold each month

Step 3   Reread and write an equation.

$R = px$ $C = f + vx$
$R = 45x$ $C = 1155 + 12x$

$$\text{Revenue} = \text{Cost}$$
$$45x = 1155 + 12x$$

Step 4   Solve the equation.

$$45x = 1155 + 12x$$
$$33x = 1155$$
$$x = 35$$

Sun Electronics must produce and sell 35 calculators per month to break even.

Step 5   Check.

The revenue for 35 calculators is 35(\$45), or \$1575. The cost for 35 calculators is \$1155 + 35(\$12) = \$1155 + \$420 = \$1575. The revenue equals the cost. It checks.

Compound Interest Problems

This type of finance problem deals with investments or loans with interest that is compounded over some time period. The compounding can be calculated annually, semiannually, quarterly, monthly, or daily.

The following formula is used to calculate compound interest:

$$A = P\left(1 + \frac{r}{n}\right)^{nt}$$

where A = the amount after some time period
P = the principal
r = the annual rate
n = the number of compounding periods per year
t = the number of years

This type of problem would be found in an intermediate algebra course.

EXAMPLE 7   **Luis invested \$1500 in an account that paid 4% annual interest compounded semiannually. Find the amount he had after 5 years.**

Solution

Step 1   Read and write givens and unknowns.

We want to find the amount, A, that Luis had after 5 years. We know that the annual interest rate, r, is 4% and the principal, P, is \$1500. We also know that the interest is compounded semiannually, or twice a year; therefore, n is 2. Also, the money is invested for 5 years, which means that t is 5.

Step 2   Begin with "Let x = the unknown."

Let A = the amount Luis will have after 5 years

Step 3   Reread and write an equation.

$$A = P\left(1 + \frac{r}{n}\right)^{nt}$$

$$= 1500\left(1 + \frac{0.04}{2}\right)^{2(5)}$$

Step 4   Solve the equation.

$$A = 1500\left(1 + \frac{0.04}{2}\right)^{2(5)}$$

$$= 1500(1 + 0.02)^{10}$$

$$= 1500(1.02)^{10}$$

$$= 1500(1.21899442)$$

$$= \$1828.49163$$

Luis had \$1828.49 in his account after 5 years.

Step 5 Check.

Using the compound interest formula (above), we get $1828.49. It checks.

PROBLEM SET 6

Solve each of the following finance problems. Be sure to follow the five-step strategy. Make a table for the simple interest problems.

Simple Interest Problems

1. Julie inherited $10,000 from her uncle. She invested part in stocks and the rest in bonds. The stocks paid an annual interest rate of 3% and the bonds paid 5% annually. If her total annual income from these two investments was $413, how much did she invest in each?

2. John deposited $2500 into two accounts at his bank. He put part of it in a platinum checking account that paid 2% annual interest. He put the rest in a savings account that paid 3% annual interest. If his total annual interest from the two accounts was $66, how much did he invest in each?

3. Dana received $1000 as a graduation gift from her grandmother. She put part of it into her savings account that pays 2% interest annually and the rest into a bond that pays 3% annually. If her total annual interest was $23.60, how much did she invest in each?

4. Mr. Lincoln invested $5000 in stocks and bonds. The stocks paid an annual interest rate of 2% and the bonds paid 5%. If his total annual interest from these two investments was $199, how much did he invest in each?

5. Joe invested $10,000 in two accounts. He put part of it in an account that paid 5% annual interest and the remainder in bonds that paid 9% annual interest. How much did he invest in each if his annual income from the two investments was $660?

6. Mrs. Ortiz inherited two different investments whose yearly income was $2100. The total value of the investments was $40,000. One was paying 4% annual interest and the other one paid 6% per year. What was the value of each investment?

7. Katie invested $50,000, part at 6% and part at 8%. The annual income on the 6% investment was $480 more than the income from the 8% investment. How much was invested at each rate?

8. Robbie invested $20,000 in stocks and bonds. The stocks paid 6% and the bonds paid 8%. The return on the stocks was $80 per year more than the return on the bonds. How much did Robbie invest in each?

9. The Smith School PTA wants to put a sum of money in an investment so that the annual interest pays for a yearly $500 scholarship for a deserving graduate. If the investment pays 5% simple interest, how much does the PTA have to invest?

10. Paul Brown wants to invest a sum of money in an investment so that the annual interest will pay his daughter's yearly tuition of $1500. If the investment pays 4% simple interest, how much does Mr. Brown have to invest?

11. How long will it take an investment of $500 to triple if it is invested at 5% simple interest?

12. How long will it take an investment of $1000 to double if it is invested at 5% simple interest?

13. Peter wants to borrow some money from his father at 4% simple interest and from his uncle at 5% simple interest. He is going to borrow $100 more from his father than from his uncle. If his total annual interest is to be $44.50, how much should he borrow from each?

14. Ross invested twice as much in an account that pays 4% simple interest as he did in an account that pays 2%. If his total annual interest is $42, how much did he invest in each account?

15. Jennifer invested three times as much in an account that pays 5% simple interest as she did in an account that pays 2%. If her total annual interest is $35.70, how much did she invest in each account?

16. Julie invests $6000 in two accounts; one pays 5% simple interest and the other pays 3%. If the interest earned on the 3% investment is $60 more than the interest earned on the 5% investment, how much is invested at each rate?

17. Patrick invests $8000 in two accounts. If the interest earned on the 4% account is $70 more than the interest earned on the 6% account, how much is invested in each account?

18. Scott received $5000 as a graduation gift from his aunt and uncle. He invested the money in stocks and bonds. This year he received a 3% annual return on the bonds, but the stocks lost 2% of their original value. If his total annual income from the stocks and the bonds was $80, how much did he invest in each?

19. Mr. Gomez received a lump sum distribution of $10,000 as a retirement incentive. He invested this money in a certificate of deposit (CD) and a technology stock. This year he received a $2\frac{1}{2}$% return on the CD, but the stock lost $1\frac{1}{2}$% on his investment. If his total annual return from these two investments was $178, how much did he invest in each?

20. The Morrows are saving for a new car. This year they invested $5000 in stocks and bonds. The bonds had an annual return of 3%, but the stocks lost 1% of their face value. If the annual return from the two investments was $82, how much did the Morrows invest in each?

Break-Even Problems

21. One factory of Gold's Optical Company manufactures a particular type of sunglasses. It costs $22 for parts and labor for each pair of sunglasses. The fixed costs are $4429 per month. If they sell these sunglasses for $65 per pair, how many pairs of sunglasses do they have to produce and sell each month to break even?

22. The Alliance Company manufactures a keychain–bottle opener that sells for $8. It costs the company $1.50 in parts and labor to produce each one, and the fixed costs are $3315 each month. How many keychain–bottle openers do they have to produce and sell each month to break even?

23. Go Surf Company manufactures a waterproof watch that sells for $140. It costs the company $35 in parts and labor to manufacture each watch. The company's fixed costs are $4515 per week. How many watches do they have to produce and sell each week to break even?

24. Time and Treasures manufactures a travel clock that sells for $65 each. It costs the company $19 in parts and labor to produce each clock, and the fixed monthly costs are $3818. How many travel clocks do they have to produce and sell each month to break even?

The following problems require algebraic techniques from intermediate algebra to solve.

Compound Interest Problems

25. Mrs. Benfield invested $2500 in an account that pays 3% annual interest compounded quarterly. Find the amount she had in the account after 3 years.

26. Eduardo received $3200 in graduation gifts. He deposited the money in an account that pays $2\frac{1}{2}$% annual interest compounded monthly. How much will he have in this account after 4 years, assuming he doesn't make any withdrawals or deposits?

27. Kathleen invested $500 in an account that pays 4% annual interest compounded monthly. Find the amount she will have in this account if she leaves the money in the bank for 6 years.

28. Alfonso invested $1800 in an account that pays $3\frac{1}{2}$% annual interest compounded semiannually. Find the amount he will have in this account after 5 years.

Motion Problems

Several types of motion problems are included in this chapter. In general, motion problems involve trains, cars, boats, planes, bicycles, or people traveling in a particular direction at a constant rate of speed for a period of time. In all the problems, we will use this formula: distance equals rate multiplied by time:

$$\boxed{\text{Distance} = \text{Rate} \cdot \text{Time}}$$

To solve the first four examples of motion problems, it will be helpful to draw a diagram to represent the distances traveled. In many cases, we will write the equation based on the relationship between the distances. We will also make a chart to organize the givens and unknowns as we have in past chapters.

Opposite-Direction Problems

The first type of motion problem we will solve involves two trains traveling in opposite directions.

EXAMPLE 1 **Two trains leave Chicago, one headed due east at 56 miles per hour and the other headed due west at 75 miles per hour. In how many hours will they be 1048 miles apart?**

Solution

Step 1 Read and write givens and unknowns.

We want to find the number of hours that each train travels until 1048 miles are between them. We know that the eastbound train's rate is 56 miles per hour and the westbound train's rate is 75 miles per hour.

Step 2 Begin with "Let x = the unknown."

Let x = the number of hours each train travels

We can draw a diagram to see the relationship between the distances:

$$
\begin{array}{c}
\text{Chicago}\\
\text{W} \longleftarrow\!\!\!\!\!\overset{}{\underset{\underset{\text{Westbound train}}{}\quad\bullet\quad\underset{\text{Eastbound train}}{}}{}}\!\!\!\!\!\longrightarrow \text{E}
\end{array}
$$

We can organize the givens and unknowns by making a table.

	Rate	Time	Distance
Eastbound train	56	x	$56x$
Westbound train	75	x	$75x$

Step 3 Reread and write an equation.

From the diagram, we can see that the distance traveled by the eastbound train plus the distance traveled by the westbound train is 1048 miles.

Distance of eastbound train + Distance of westbound train = 1048 miles

$$56x \qquad\qquad + \qquad\qquad 75x \qquad\qquad = 1048$$

Step 4 Solve the equation.

$$56x + 75x = 1048$$
$$131x = 1048$$
$$x = 8$$

After 8 hours, the two trains will be 1048 miles apart.

Step 5 Check.

The distance traveled by the eastbound train is 56(8) or 448 miles.
The distance traveled by the westbound train is 75(8) or 600 miles.
The trains will be 448 + 600 or 1048 miles apart. It checks.

Pursuit Problems

The second type of motion problem we will solve involves a moped pursuing a bicycle along the same route.

EXAMPLE 2 **A bicyclist leaves Morro Bay traveling at 12 miles per hour along Highway 1. Half an hour later a friend on a moped leaves from the same location in Morro Bay pursuing the bicyclist at a speed of 20 miles per hour. How long will it take the friend on the moped to catch up with the bicyclist?**

Solution

Step 1   Read and write givens and unknowns.

We want to find the number of hours traveled by the friend on the moped. We know that the bicycle's rate is 12 miles per hour and the moped's rate is 20 miles per hour. We also know that the time of the bicyclist is $\frac{1}{2}$ hour more than the time of the friend on the moped.

Step 2   Begin with "Let x = the unknown."

Let x = the number of hours moped has traveled

$x + \frac{1}{2}$ = the number of hours bicycle has traveled

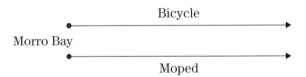

	Rate	**Time**	**Distance**
Bicycle	12	$x + \frac{1}{2}$	$12\left(x + \frac{1}{2}\right)$
Moped	20	x	$20x$

Step 3   Reread and write an equation.

From the diagram, we can see that the distances are equal.

Distance of moped = Distance of bicycle

$$20x = 12\left(x + \frac{1}{2}\right)$$

Step 4   Solve the equation.

$$20x = 12\left(x + \frac{1}{2} \right)$$

$$20x = 12x + 6$$

$$8x = 6 \qquad\qquad x + \frac{1}{2} = 0.75 + 0.5$$

$$x = 0.75 \qquad\qquad\qquad = 1.25$$

It will take 0.75 hour or 45 minutes for the friend on the moped to catch up with the bicyclist.

Step 5   Check.

The bicyclist has traveled for 1.25 hours at 12 miles per hour, or $12(1.25) = 15$ miles. The friend on the moped has traveled 0.75 hour at 20 miles per hour, or $20(0.75) = 15$ miles. Therefore, the distances are equal and it checks.

Round-Trip Problems

The third type of motion problem we will solve involves a motorboat that is making a trip from the river delta to the bay and then returning along the same route.

EXAMPLE 3   **A motorboat traveled from the river delta to the bay at a speed of 50 miles per hour and returned at a speed of 30 miles per hour. If the round trip took 6 hours, how long did it take to get from the river delta to the bay?**

Solution

Step 1   Read and write givens and unknowns.

We want to find the number of hours the motorboat traveled going from the river delta to the bay. We know that the rate going was 50 miles per hour and the rate returning was 30 miles per hour. We also know that the total time was 6 hours.

Step 2   Begin with "Let x = the unknown."

Let x = the number of hours going
$6 - x$ = the number of hours returning

	Rate	Time	Distance
Going	50	x	$50x$
Returning	30	$6 - x$	$30(6 - x)$

Step 3 Reread and write an equation.

We can see from the diagram that the distances are equal.

Distance going = Distance returning
$$50x = 30(6 - x)$$

Step 4 Solve the equation.

$$50x = 30(6 - x)$$
$$50x = 180 - 30x$$
$$80x = 180$$
$$x = 2.25 \qquad 6 - x = 6 - 2.25 = 3.75$$

It will take the motorboat 2.25 hours to get from the river delta to the bay, and it will take 3.75 hours to return.

Step 5 Check.

The distance going is 50(2.25) or 112.5 miles. The distance returning is 30(3.75) or 112.5 miles. The distances are equal. It checks.

Same-Direction Problems

The fourth type of motion problem we will solve involves two people leaving the same place at the same time and traveling in the same direction at different rates.

EXAMPLE 4 **Samantha and Julie leave the office at the same time and travel in the same direction along parallel roads. Samantha's speed is nine less than four times Julie's speed. Three hours later Samantha is 162 miles ahead of Julie. Find the speeds of each.**

Solution

Step 1   Read and write givens and unknowns.

We want to find Samantha's and Julie's rates. We know that Samantha's rate of speed is nine less than four times Julie's rate of speed. We also know that after they have traveled for 3 hours, the distance between them is 162 miles. In other words, Samantha's distance minus Julie's distance is 162 miles.

Step 2   Begin with "Let x = the unknown."

Let x = Julie's rate
$4x - 9$ = Samantha's rate

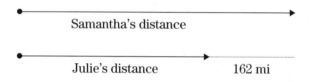

	Rate	**Time**	**Distance**
Julie	x	3	$3x$
Samantha	$4x - 9$	3	$3(4x - 9)$

Step 3   Reread and write an equation.

From the diagram, we can see that Samantha's distance minus Julie's distance is 162 miles.

Samantha's distance $-$ Julie's distance = 162 miles
$$3(4x - 9) \quad - \quad 3x \quad = \quad 162$$

Step 4   Solve the equation.

$$3(4x - 9) - 3x = 162$$
$$12x - 27 - 3x = 162$$
$$9x - 27 = 162$$
$$9x = 189$$
$$x = 21 \qquad 4x - 9 = 4(21) - 9 = 84 - 9 = 75$$

Julie is traveling at 21 miles per hour, and Samantha is traveling at 75 miles per hour.

Step 5 Check.

After 3 hours, Julie traveled 21(3) or 63 miles and Samantha traveled 75(3) or 225 miles. They are 225 − 63 or 162 miles apart at that time. It checks.

The next two examples are more difficult and are usually found in intermediate algebra courses.

Problems Involving a Current

The fifth type of motion problem involves a boat traveling with the current and against the current. When it travels with the current, its speed is the sum of its speed in still water and the speed of the current. When it travels against the current, its speed is its speed in still water minus the speed of the current.

Speed with the current	=	Speed in still water	+	Speed of the current
Speed against the current	=	Speed in still water	−	Speed of the current

EXAMPLE 5 **A boat travels 129 miles with the current in the same time it travels 99 miles against the current. If the speed of the current is 5 miles per hour, find the speed of the boat in still water.**

Solution

Step 1 Read and write givens and unknowns.

We want to find the speed of the boat in still water. We know that it can travel a distance of 129 miles with the current in the same time it can travel 99 miles against the current.

Step 2 Begin with "Let x = the unknown."

Let x = the rate (or speed) of the boat in still water
$x + 5$ = the rate of the boat with the current
$x - 5$ = the rate of the boat against the current

	Rate	Time	Distance
With the current	$x + 5$		129
Against the current	$x - 5$		99

We want to find the time given the rate and the distance. We know that the rate multiplied by the time equals the distance. Therefore, the time is equal to the distance divided by the rate.

$$\text{Rate} \cdot \text{Time} = \text{Distance}$$
$$\text{Time} = \frac{\text{Distance}}{\text{Rate}}$$

	Rate	Time	Distance
With the current	$x + 5$	$\dfrac{129}{x + 5}$	129
Against the current	$x - 5$	$\dfrac{99}{x - 5}$	99

Step 3 Reread and write an equation.

The time with the current = the time against the current

$$\frac{129}{x + 5} = \frac{99}{x - 5}$$

Step 4 Solve the equation.

$$\frac{129}{x + 5} = \frac{99}{x - 5}$$
$$129(x - 5) = 99(x + 5)$$
$$129x - 645 = 99x + 495$$
$$30x = 1140$$
$$x = 38$$

The speed of the boat in still water is 38 miles per hour.

Step 5 Check.

The rate with the current is $38 + 5$ or 43 miles per hour. To travel 129 miles, it would take $129 \div 43$ or 3 hours. The rate against the current is $38 - 5$ or 33 miles per hour. To travel 99 miles, it would take $99 \div 33$ or 3 hours. The times are the same. It checks.

To solve the next example, we must use a system of two equations in two variables.

EXAMPLE 6 **Tim rows his boat 36 miles upstream in 3 hours. It takes him 2 hours to return downstream. Find his speed in still water and the speed of the current.**

Solution

Step 1 Read and write givens and unknowns.

We want to find the speed of the boat in still water and the speed of the current. We know that Tim goes 36 miles in 3 hours upstream (against the current) and he goes 36 miles in 2 hours downstream (with the current).

Step 2 Begin with "Let x = the unknown."

Let x = the rate (or speed) of the boat in still water
y = the rate of the current
$x + y$ = the rate of the boat with the current
$x - y$ = the rate of the boat against the current

	Rate	Time	Distance
With the current	$x + y$	2	$2(x + y)$
Against the current	$x - y$	3	$3(x - y)$

Step 3 Reread and write an equation.

The distance each way is 36 miles.

$2(x + y) = 36$
$3(x - y) = 36$

Step 4 Solve the system of equations.

$2x + 2y = 36 \qquad 3(2x + 2y) = 3(36)$
$3x - 3y = 36 \qquad -2(3x - 3y) = -2(36)$

$$
\begin{aligned}
6x + 6y &= 108 \\
-6x + 6y &= -72 \\
\hline
12y &= 36 \\
y &= 3
\end{aligned}
\qquad
\begin{aligned}
2x + 2y &= 36 \\
2x + 6 &= 36 \\
2x &= 30 \\
x &= 15
\end{aligned}
$$

The speed of the boat in still water is 15 miles per hour, and the speed of the current is 3 miles per hour.

Step 5   Check.

The speed against the current is $15 - 3$ or 12 miles per hour. In 3 hours, the boat travels 12(3) or 36 miles upstream. The speed with the current is $15 + 3$ or 18 miles per hour. In 2 hours, the boat travels 18(2) or 36 miles downstream. It checks.

PROBLEM SET 7

Solve each of the following motion problems. Be sure to follow the five-step strategy. Make a table and a diagram for each problem.

1. Two private jets start from Chicago and travel in opposite directions. The speed of the first jet is ten less than two times the speed of the second jet. In 3 hours they are 1050 miles apart. Find the speed of each jet. (Opposite direction)

2. Two trains leave Springfield at the same time traveling in opposite directions. The eastbound train travels at 48 miles per hour, and the westbound train travels at 54 miles per hour. In how many hours will they be 255 miles apart? (Opposite direction)

3. Two search and rescue teams leave camp at the same time and walk in opposite directions looking for a lost child. The speed of the first team is three less than twice the speed of the second team. In 4 hours they are 36 miles apart. Find the speed of each team. (Opposite direction)

4. A sailboat leaves the harbor traveling at 36 miles per hour. Four hours later a motorboat begins to pursue the sailboat at a speed of 60 miles per hour. How long will it take the motorboat to catch up to the sailboat? How far from the harbor will they meet? (Pursuit)

5. A freight train starts from Los Angeles and heads for Seattle at 30 miles per hour. Two hours later a passenger train leaves the same station for Seattle traveling at 50 miles per hour. How long will it take the passenger train to pass the freight train? (Pursuit)

6. A tour bus starts from Memphis and heads for Miami at 50 miles per hour. Three hours later a truck pursues the bus at a speed of 65 miles per hour. How long will it take the truck to catch up to the bus? (Pursuit)

7. A ferry travels from the dock to a cruise ship at a speed of 40 miles per hour and returns at a speed of 25 miles per hour. If the entire trip took 13 hours, how long did it take the ferry to travel from the dock to the cruise ship? (Round-trip)

8. A helicopter traveled from the airport to the site of a fire at a speed of 60 miles per hour and returned at a speed of 40 miles per hour. If the entire trip took 3 hours, how long did it take the helicopter to get to the site of the fire? (Round-trip)

9. It takes a sailboat a total of 2 hours to travel from the harbor to an island and return. The sailboat travels to the island at 30 miles per hour and returns at a rate of 20 miles per hour. What is the distance from the harbor to the island? (Round-trip)

10. Five hours after Jeff and Jim leave the mall at the same time and head in the same direction, Jim is 235 miles ahead of Jeff. If Jim's speed is 1 mile per hour less than three times Jeff's speed, find the speeds of Jeff and Jim. (Same direction)

11. Two trains leave the Santa Barbara station at the same time traveling on parallel tracks. The express train travels at 75 miles per hour, and the local train travels at 50 miles per hour. In how many hours will they be 185 miles apart? (Same direction)

12. Judith and Danya leave their house and travel in the same direction. Danya's speed is 1 mile per hour more than two times Judith's speed. After 3 hours Danya is 15 miles ahead of Judith. Find their speeds. (Same direction)

13. Barbara can paddle her canoe 48 miles with the current in the same time it takes her to paddle 24 miles against the current. If the speed of the current is 2 miles per hour, find the speed of the boat in still water. (Current)

14. Jim can fly his airplane 3825 kilometers with the wind in the same time it takes him to fly 3675 kilometers against the wind. If the speed of the wind is 10 kilometers per hour, find the speed of the airplane in still air. (Current)

15. Scott rows his boat 70 kilometers downstream in 3.5 hours. It takes him 5 hours to return upstream. Find his speed in still water and the speed of the current. (Use a system of equations to solve this current problem.)

16. Sharon can bike 70 miles with the wind in 5 hours. It takes her 7 hours to return against the wind. Find her speed in still air and the speed of the wind. (Use a system of equations to solve this current problem.)

The following problems are not labeled with their type. Decide which type of motion problem each one is and then solve it.

17. The Amtrak passenger train leaves Washington, DC, and heads north at 50 miles per hour. Two hours later the Metroliner leaves the same station heading north traveling at 60 miles per hour. How long will it take the Metroliner to catch up to the Amtrak passenger train?

18. Frank drives from his office to the corporate headquarters at 60 miles per hour. Because of traffic, he returns at 45 miles per hour. If the entire trip took 7 hours, how far is it from his office to corporate headquarters?

19. Two cars leave Dallas, one headed north at 50 miles per hour and the other headed south at 36 miles per hour. In how many hours will they be 301 miles apart?

20. Rachel and Ann leave their apartment at the same time and walk north along the same route. Rachel's speed is three less than twice Ann's speed. Two hours later Rachel is 2 miles ahead of Ann. Find the speed of each.

21. Monica can cross-country ski 18 miles with the wind in 2 hours. It takes her 3 hours to return against the wind. Find the speed of the wind and her speed in still air. (This problem requires a system of equations.)

22. Diane and Pat are part of a group on a bicycle tour. Diane can ride at a rate of 12 miles per hour, and Pat can ride at 9 miles per hour. If they both leave their hotel at the same time and ride in the same direction, how long will it take them to be 15 miles apart?

23. An air traffic controller has noticed two planes that are 1350 miles apart approaching each other, one from the east and one from the west. The eastbound plane is traveling at 400 miles per hour, and the westbound plane is traveling at 500 miles per hour. How long does the controller have to alter one plane's course so that the planes do not collide?

24. Heather can paddle her kayak 33 miles with the current in the same time it takes her to paddle 15 miles against the current. If the speed of the current is 3 miles per hour, find her speed in still water.

25. An air ambulance traveled from the hospital to the accident site and back in 35 minutes. Its speed going was 200 miles per hour, and its speed returning was 150 miles per hour. How long did it take the air ambulance to arrive at the scene of the accident? (*Hint:* Change 35 minutes to hours before working the problem.)

8

Work Problems

Work problems involve people or machines doing some job together. This type of problem is also known as shared-work problems. They might involve three people roofing a house together or two faucets filling a pool together.

To solve work problems the units (hours, minutes, days, etc.) within each problem must be the same. If they are not the same, then we must convert the units before we begin the problem.

We always need to find the *rate of work*, which is the fractional part of the job that can be done in one unit of time. For example, if Jim can paint a barn in 4 days, he can paint $\frac{1}{4}$ of the barn each day. Therefore, his rate of work per day is $\frac{1}{4}$. To calculate the rate of work, we divide 1 by the time it takes to do the job alone:

$$\text{Rate of work} = \frac{1}{\text{Time to do job alone}}$$

In work problems, the rate of work times the time worked equals the fractional part of the job done by each entity. In our example, if Jim works for 3 days, he can paint $\frac{1}{4}(3)$ or $\frac{3}{4}$ of the barn.

$$\text{Rate of work} \cdot \text{Time worked} = \text{Fractional part of job done}$$

To set up the equation for these problems, we use the principle that the total of the fractional parts equals one whole job. The only exception to this is if we want to complete only part of the job.

$$\begin{array}{c} \text{Fractional part} \\ \text{done by \#1} \end{array} + \begin{array}{c} \text{Fractional part} \\ \text{done by \#2} \end{array} = \text{One whole job}$$

EXAMPLE 1 **Jim can paint a barn in 4 days. Bill can paint the same barn in 3 days. How long would it take them to paint the barn together?**

Solution

Step 1 Read and write givens and unknowns.

We want to find the number of days it would take Jim and Bill to paint the barn together. We know that Jim's time alone is 4 days and Bill's time alone is 3 days. We can find each rate using the time alone.

Step 2 Begin with "Let x = the unknown."

Let x = the number of hours to paint barn together
(*Note:* This is the same as the time worked.)

We can organize the givens and unknowns by making a table.

	Rate of work	Time worked	Fractional part of job done
Jim	$\frac{1}{4}$	x	$\frac{1}{4}x$
Bill	$\frac{1}{3}$	x	$\frac{1}{3}x$

Step 3 Reread and write an equation.

$$\begin{array}{c} \text{Fractional part} \\ \text{done by Jim} \end{array} + \begin{array}{c} \text{Fractional part} \\ \text{done by Bill} \end{array} = \text{One whole job}$$

$$\frac{1}{4}x \quad + \quad \frac{1}{3}x \quad = \quad 1$$

Step 4 Solve the equation.

$$\frac{1}{4}x + \frac{1}{3}x = 1$$

$$12\left(\frac{1}{4}x\right) + 12\left(\frac{1}{3}x\right) = 12(1)$$

$$3x + 4x = 12$$

$$7x = 12$$

$$x = \frac{12}{7} = 1\frac{5}{7}$$

It would take $1\frac{5}{7}$ days for Jim and Bill to paint the barn together.

Step 5 Check.

In $1\frac{5}{7}$ days, Jim will paint $\frac{1}{4}\left(\frac{12}{7}\right) = \frac{3}{7}$ of the barn. In $1\frac{5}{7}$ days, Bill will paint $\frac{1}{3}\left(\frac{12}{7}\right) = \frac{4}{7}$ of the barn. Together they will do $\frac{3}{7} + \frac{4}{7} = 1$ whole job. It checks.

EXAMPLE 2 **Helene can deliver papers on a particular route in 5 hours. Together she and Joanne can deliver papers on the same route in 2 hours. How long would it take Joanne to deliver the papers alone?**

Solution

Step 1 Read and write givens and unknowns.

We want to find the number of hours it would take Joanne to deliver the papers alone. We know that Helene's individual time is 5 hours. We also know that their time worked together is 2 hours.

Step 2 Begin with "Let x = the unknown."

Let x = the number of hours for Joanne to deliver the papers alone

$\frac{1}{x}$ = Joanne's rate of work per hour

	Rate of work	Time worked	Fractional part of job done
Helene	$\dfrac{1}{5}$	2	$\dfrac{1}{5}\left(\dfrac{2}{1}\right)=\dfrac{2}{5}$
Joanne	$\dfrac{1}{x}$	2	$\dfrac{1}{x}\left(\dfrac{2}{1}\right)=\dfrac{2}{x}$

Step 3 Reread and write an equation.

$$\begin{array}{ccccc} \text{Fractional part} & & \text{Fractional part} & & \\ \text{done by Helene} & + & \text{done by Joanne} & = & \text{One whole job} \\ \dfrac{2}{5} & + & \dfrac{2}{x} & = & 1 \end{array}$$

Step 4 Solve the equation.

$$\frac{2}{5}+\frac{2}{x}=1$$

$$5x\left(\frac{2}{5}\right)+5x\left(\frac{2}{x}\right)=5x(1)$$

$$2x+10=5x$$

$$10=3x$$

$$\frac{10}{3}=x$$

It would take Joanne $\dfrac{10}{3}$ or $3\dfrac{1}{3}$ hours to deliver the papers alone.

Step 5 Check.

In 2 days, Helene will deliver $\dfrac{1}{5}(2)$ or $\dfrac{2}{5}$ of the papers. Joanne's rate is $1\div\dfrac{10}{3}=1\left(\dfrac{3}{10}\right)=\dfrac{3}{10}$. Joanne will deliver $\dfrac{3}{10}\left(\dfrac{2}{1}\right)=\dfrac{3}{5}$ of the papers. Together, they will do $\dfrac{2}{5}+\dfrac{3}{5}=1$ whole job.

The next examples are more difficult and might be seen in an intermediate algebra course.

EXAMPLE 3 **Peter can paint a room in 8 hours. George can paint the same room in 6 hours. They start to paint the room to-**

gether. **After 2 hours Peter leaves for lunch and George finishes the job alone. How long does it take George to finish?**

Solution

Step 1 Read and write givens and unknowns.

We want to find the number of hours it takes George to finish painting the room. We know that Peter's time alone is 8 hours and George's time alone is 6 hours. These will give us their individual rates. We also know that Peter works only 2 hours and George works 2 hours plus the time it takes him to finish the job.

Step 2 Begin with "Let x = the unknown."

Let x = the number of hours for George to finish painting the room
$x + 2$ = the number of hours George works

	Rate of work	Time worked	Fractional part of job done
George	$\dfrac{1}{6}$	$x + 2$	$\dfrac{1}{6}\left(\dfrac{x+2}{1}\right) = \dfrac{x+2}{6}$
Peter	$\dfrac{1}{8}$	2	$\dfrac{1}{8}\left(\dfrac{2}{1}\right) = \dfrac{1}{4}$

Step 3 Reread and write an equation.

$$\begin{matrix} \text{Fractional part} \\ \text{done by George} \end{matrix} + \begin{matrix} \text{Fractional part} \\ \text{done by Peter} \end{matrix} = \text{One whole job}$$

$$\dfrac{x+2}{6} \quad + \quad \dfrac{1}{4} \quad = \quad 1$$

Step 4 Solve the equation.

$$\dfrac{x+2}{6} + \dfrac{1}{4} = 1$$

$$12\left(\dfrac{x+2}{6}\right) + 12\left(\dfrac{1}{4}\right) = 12(1)$$

$$2(x+2) + 3 = 12$$

$$2x + 4 + 3 = 12$$

$$2x + 7 = 12$$

$$2x = 5$$

$$x = \dfrac{5}{2} = 2\dfrac{1}{2}$$

It will take George $2\frac{1}{2}$ hours to finish painting the room. (Peter had a long lunch!)

Step 5 Check.

In 2 hours, Peter will paint $\frac{1}{8}(2) = \frac{1}{4}$ of the room. George works for $2 + 2\frac{1}{2} = 4\frac{1}{2} = \frac{9}{2}$ hours. George will paint $\frac{1}{6}\left(\frac{9}{2}\right) = \frac{3}{4}$ of the room. Together they will do $\frac{1}{4} + \frac{3}{4} = 1$ whole job. It checks.

Example 4 involves two actions working against each other: filling a pool and emptying the pool. In this case we represent the opposing action with a negative number. Because we are filling the pool, the opposing action is emptying the pool. The rate of work done by the drain will be represented with a negative number.

EXAMPLE 4 **A swimming pool can be filled in 12 hours and emptied in 15 hours. One day while the pool is being filled, the drain is accidentally left open. It is discovered after 3 hours and the drain is closed. How much longer will it take to fill the pool?**

Solution

Step 1 Read and write givens and unknowns.

We want to find the number of hours it takes to finish filling the pool after the drain is closed. We know that it takes 12 hours to fill the pool alone and 15 hours to drain the pool alone. We also know that the drain works for only 3 hours. Because filling and draining are opposite actions, we will represent the rate of draining with a negative number.

Step 2 Begin with "Let x = the unknown."

Let x = the number of hours it takes to finish filling the pool
$x + 3$ = the number of hours to fill the pool

	Rate of work	Time worked	Fractional part of job done
Filling	$\dfrac{1}{12}$	$x + 3$	$\dfrac{1}{12}\left(\dfrac{x+3}{1}\right) = \dfrac{x+3}{12}$
Draining	$-\dfrac{1}{15}$	3	$-\dfrac{1}{15}\left(\dfrac{3}{1}\right) = -\dfrac{1}{5}$

Step 3 Reread and write an equation.

Fractional part filling + Fractional part draining = One whole job

$$\frac{x+3}{12} \qquad + \qquad \left(-\frac{1}{5}\right) \qquad = \qquad 1$$

Step 4 Solve the equation.

$$\frac{x+3}{12} - \frac{1}{5} = 1$$

$$60\left(\frac{x+3}{12}\right) - 60\left(\frac{1}{5}\right) = 60(1)$$

$$5(x+3) - 12 = 60$$

$$5x + 15 - 12 = 60$$

$$5x + 3 = 60$$

$$5x = 57$$

$$x = \frac{57}{5} = 11\frac{2}{5}$$

It will take $11\dfrac{2}{5}$ hours to finish filling the pool after the drain is closed.

Step 5 Check.

The faucet works $14\dfrac{2}{5}$ or $\dfrac{72}{5}$ hours at a rate of $\dfrac{1}{12}$ of the pool per hour. The fractional part done is $\dfrac{72}{5}\left(\dfrac{1}{12}\right) = \dfrac{6}{5} = 1\dfrac{1}{5}$ of the job. The drain works for 3 hours at a rate of $-\dfrac{1}{15}$ of the pool per hour. The fractional part done is $3\left(-\dfrac{1}{15}\right) = -\dfrac{1}{5}$ of the job. Since $1\dfrac{1}{5} - \dfrac{1}{5} = 1$ whole job, it checks.

PROBLEM SET 8

Solve each of the following work problems. Be sure to follow the five-step strategy. Make a table for each problem.

1. It takes Ken 15 hours to paint a fence alone. Max can paint the same fence in 10 hours alone. How long would it take them to paint the fence together?

2. Jeff and George work at the Golden Tee Motel cleaning rooms. Jeff can clean all the rooms by himself in 8 hours. George can clean all the rooms by himself in 6 hours. If they work together, how long would it take them to clean all the rooms?

3. Jerry can panel a room by himself in 3 days, and Jose can do the same job in 2 days. How long would it take them to panel the room together?

4. Matt can mow a lawn alone in 30 minutes. His assistant takes 45 minutes to do the same job. How long would it take them to mow the lawn together?

5. Susan can wash a car in 20 minutes, and it takes Emily 30 minutes to wash the same car. How long would it take them to wash the car together?

6. Tom, John, and Bill have volunteered to fence the preschool playground. If it would take Tom 4 days to build the fence alone, John 3 days, and Bill 5 days, how long would it take them working together?

7. Eleanor and Sheila are working on a project stuffing envelopes for a fundraiser. It would take Eleanor 25 hours to stuff all the envelopes alone. She and Sheila can do the job in 15 hours. How long would it take Sheila to do the job alone?

8. Jim and Robbie can weed their garden together in 5 hours. If Jim is working alone, it would take him 8 hours to weed the garden. How long would it take Robbie to weed the garden alone?

9. Tim and Dale can tune up a car together in 2 hours. If Tim is working alone, it takes him 3 hours. How long would it take Dale to tune up the car alone?

10. Shane and Aaron can deliver papers on a particular route in 3 hours. If Shane is working alone, it takes him 5 hours to deliver the papers. How long would it take Aaron to deliver the papers alone?

11. Two pumps can fill a tank in 6 hours working together. It takes the first pump 10 hours to fill the tank alone. How long would it take the second pump to fill the tank alone?

12. It takes Joel 20 hours to overhaul a particular engine. Together he and Mario can overhaul that engine in 12 hours. How long would it take Mario to do this job alone?

13. A butcher can grind a package of meat in 20 seconds. It takes his assistant 30 seconds to do the same job. How long would it take them to do this job together?

14. It takes a machine $3\frac{1}{2}$ hours to do a particular job. An older machine can do the same job in 5 hours. How long would it take the two machines to do the job together?

15. Two drains can empty a pool together in 6 hours. It takes the first drain 9 hours to drain the pool alone. How long would it take the second drain to do the job alone?

16. Town Center Supplies has two copiers. Working together, the copiers can do a certain job in 20 minutes. It would take the faster copier 30 minutes to do this job alone. How long would it take the slower copier to do this job alone?

17. Hank can mow the lawn alone in 1 hour and 30 minutes. It takes his son 2 hours to do the same job alone. How long would it take them to mow the lawn together? (*Caution*: Convert Hank's time alone to hours first.)

18. Julia can wash the family car in 45 minutes. It takes her daughter 1 hour to do the same job. How long would it take them to wash the car together? (*Caution*: Convert the daughter's time alone to minutes first.)

The following problems are more difficult and are from intermediate algebra.

19. Harry can wash the car in 25 minutes, and it takes Sheryl 30 minutes to wash the car. They start washing the car together. After 10 minutes, the phone rings and Harry answers it. Sheryl finishes washing the car by herself. How long does it take Sheryl to finish washing the car?

20. Margie can weed the garden in 3 hours, but it takes Grace 2 hours to weed the garden. They start weeding the garden together. After 45 minutes, Grace leaves and Margie finishes the job alone. How long does it take Margie to finish weeding the garden? (*Caution*: Convert 45 minutes to hours first.)

21. It takes 20 minutes to fill the tub using only the hot water faucet. It takes 15 minutes to fill the tub using only the cold water faucet. Susan turns both faucets on to fill the tub. After 7 minutes, she realizes that the water is too cool for her bath and turns off the cold water. How much longer will it take to fill the tub with only the hot water faucet on?

22. A tank can be filled in 10 hours and emptied in 12 hours. One day while the tank is being filled, the drain is accidentally left open. After 2 hours, the mistake is discovered and the drain is closed. How much longer does it take to fill the tank?

23. A pool can be filled in 8 hours and drained in 10 hours. One day William begins to fill the pool with the drain left open. He discovers his mistake after an hour and a half and closes the drain. How much longer does it take to fill the tank?

24. A tub can be filled in 2 hours and emptied in 3 hours. One day while the tub is being filled, the drain is accidentally left open. After $\frac{1}{2}$ hour, the mistake is discovered and the drain is closed. How much longer does it take to fill the tub?

Variation Problems

Many mathematical relationships involve variation or proportionality in which one quantity is proportional to another. The three types of variation are *direct, joint,* and *inverse.* We will look at problems involving these three types as well as combined variation problems.

Direct Variation Problems

The words "y varies directly with x" or "y is directly proportional to x" mean that $y = kx$ for some constant k. The constant k is called the *constant of variation* or the *constant of proportionality.*

EXAMPLE 1 **Express each statement as a formula, using k as the constant of variation or the constant of proportionality.**

Statement	Formula
y varies directly with the cube of t.	$y = kt^3$
P is directly proportional to the square root of x.	$P = k\sqrt{x}$
The area of a circle varies directly with the square of its radius.	$A = kr^2$

EXAMPLE 2 **Assume that y varies directly with x. If x is 2 when y is 12, find the value of y when x is 7.**

Solution

Step 1 Read and write givens and unknowns.

We want to find the value of y when x is 7. We will also need to find the value of k. We know that y varies directly with x and also that $x = 2$ when $y = 12$.

Step 2 Begin with "Let $x =$ the unknown."

Let $y =$ the value when x is 7

Step 3 Reread and write an equation.

The statement "y varies directly with x" translates to $y = kx$. We know that $x = 2$ when $y = 12$. We want to find y when $x = 7$.

Step 4 Solve the equation.

$$y = kx \qquad \text{and} \qquad y = kx$$
$$12 = k(2) \qquad\qquad\quad y = 6(7)$$
$$6 = k \qquad\qquad\qquad\quad y = 42$$

When x is 7, the value of y is 42.

Step 5 Check.

We will check the equation $y = kx$ with $k = 6$ for y is 12 when x is 2 and for y is 42 when x is 7. The statement $12 = 6(2)$ is true and the statement $42 = 6(7)$ is true. It checks.

Joint Variation Problems

The words "z varies jointly with x and y" or "z is jointly proportional to x and y" mean that $z = kxy$ for some constant k. The constant k is called the *constant of variation* or the *constant of proportionality*.

EXAMPLE 3 **Express each statement as a formula, using k as the constant of variation or the constant of proportionality.**

Statement	Formula
y varies jointly with x and the square of t.	$y = kxt^3$
P is jointly proportional to a and the cube of b.	$P = kab^3$
The area of a triangle varies jointly with its base and its height.	$A = kbh$

EXAMPLE 4 **The volume of a right circular cone is jointly proportional to the area of its base and its height. The volume of a right circular cone is 40 cubic inches when the area of its base is 20 square inches and its height is 6 inches. Find the volume of a cone when the area of its base is 33 square inches and its height is 11 inches.**

Solution

Step 1 Read and write givens and unknowns.

We want to find the volume of a right circular cone when the area of its base is 33 square inches and its height is 11 inches. We know that the volume of the cone is jointly proportional to the area of its base and its height. We also know that the volume is 40 cubic inches when the area of its base is 20 square inches and the height is 6 inches.

Step 2 Begin with "Let x = the unknown."

Let V = volume of the cone (in cubic inches)
  B = area of its base (in square inches)
  h = its height (in inches)

Step 3 Reread and write an equation.

The statement "the volume, V, is jointly proportional to the area of its base, B, and its height, h" translates to $V = kBh$. We know that $V = 40$ when $B = 20$ and $h = 6$. We want to find V when $B = 33$ and $h = 11$.

Step 4 Solve the equation.

$$V = kBh \qquad \text{and} \qquad V = kBh$$

$$40 = k(20)(6) \qquad\qquad = \frac{1}{3}(33)(11)$$

$$40 = 120k \qquad\qquad = 121$$

$$\frac{1}{3} = k$$

The volume of the right circular cone is 121 cubic inches.

Step 5 Check.

We will check the equation $V = kBh$ with $k = \dfrac{1}{3}$ for $V = 40$ when $B = 20$ and $h = 6$ and for $V = 121$ when $B = 33$ and $h = 11$. The statement $40 = \dfrac{1}{3}(20)(6)$ is true, and the statement $121 = \dfrac{1}{3}(33)(11)$ is true. It checks.

Inverse Variation Problems

The words "y varies inversely with x" or "y is inversely proportional to x" mean that $y = \dfrac{k}{x}$ for some constant k. The constant k is called the *constant of variation* or the *constant of proportionality*.

EXAMPLE 5 **Express each statement as a formula, using k as the constant of variation or the constant of proportionality.**

Statement	Formula
y varies inversely with the cube of t.	$y = \dfrac{k}{t^3}$
P is inversely proportional to the square root of x.	$P = \dfrac{k}{\sqrt{x}}$
The rate of speed a car travels a certain distance is inversely proportional to the time it takes to get there.	$r = \dfrac{k}{t}$

EXAMPLE 6 **The width of a rectangle of fixed area varies inversely with the length. If the width of the rectangle is 4 centimeters, its length is 9 centimeters. What is the length when the width is 8 centimeters?**

Solution

Step 1 Read and write givens and unknowns.

We want to find the length of the rectangle of fixed area when its width is 8 centimeters. We know that the width of a rectangle of fixed area varies inversely with the length. We also know that when the width is 4 centimeters, the length is 9 centimeters.

Step 2 Begin with "Let x = the unknown."

Let l = the length of a rectangle of fixed area (in centimeters)
    w = the width of a rectangle of fixed area (in centimeters)

Step 3 Reread and write an equation.

The statement "the width, w, of a rectangle of fixed area varies inversely with the length, l" translates to $w = \dfrac{k}{l}$. We know that $w = 4$ when $l = 9$. We want to find l when $w = 8$.

Step 4 Solve the equation.

$$w = \frac{k}{l} \quad \text{and} \quad w = \frac{k}{l}$$

$$4 = \frac{k}{9} \qquad\qquad 8 = \frac{36}{l}$$

$$36 = k \qquad\qquad 8l = 36$$

$$l = 4.5$$

When the width of a rectangle of fixed area is 8 centimeters, the length is 4.5 centimeters.

Step 5 Check.

We will check the equation $w = \dfrac{k}{l}$ with $k = 36$ for $w = 4$ when $l = 9$ and for $w = 8$ when $l = 4.5$. The statement $4 = \dfrac{36}{9}$ is true and the statement $4.5 = \dfrac{36}{8}$ is true. It checks.

Combined Variation Problems

Many applied problems involve a combination of direct, inverse, and joint variation. These problems are called combined variation. We use only one constant of variation in each problem.

EXAMPLE 7   **Express each statement as a formula, using k as the constant of variation or the constant of proportionality.**

Statement	Formula
y varies directly with x and inversely with the square of t.	$y = \dfrac{kx}{t^2}$
R is directly proportional to l and inversely proportional to the square of r.	$R = \dfrac{kl}{r^2}$
S varies jointly with x and y and inversely with the cube of z.	$S = \dfrac{kxy}{z^3}$

EXAMPLE 8   **The electrical resistance of a wire is directly proportional to its length and inversely proportional to the square of its radius. A wire with a $\dfrac{1}{2}$-inch radius and 75 inches long has a resistance of 12 ohms. Find the resistance of 90 inches of the same kind of wire with a $\dfrac{1}{4}$-inch radius.**

Solution

Step 1   Read and write givens and unknowns.

We want to find the resistance of 90 inches of wire with a $\dfrac{1}{4}$-inch radius.

We know that the resistance of a wire is directly proportional to its length and inversely proportional to the square of its radius. We also know that when the length is 75 inches and the width is $\dfrac{1}{2}$ inch, the resistance is 12 ohms.

Step 2   Begin with "Let $x =$ the unknown."

Let $R =$ the resistance of the wire (in ohms)
$r =$ the radius of the wire (in inches)
$l =$ the length of the wire (in inches)

Step 3   Reread and write an equation.

The statement "the resistance of a wire is directly proportional to its length and inversely proportional to the square of its radius" translates to $R = \dfrac{kl}{r^2}$. We also know that when $R = 12$, $l = 75$ and $r = \dfrac{1}{2}$. We want to find R when $l = 90$ and $r = \dfrac{1}{4}$.

Step 4   Solve the equation.

$$R = \frac{kl}{r^2} \qquad \text{and} \qquad R = \frac{kl}{r^2}$$

$$12 = \frac{k(75)}{\left(\dfrac{1}{2}\right)^2} \qquad\qquad = \frac{0.04(90)}{\left(\dfrac{1}{4}\right)^2}$$

$$12 = k(300) \qquad\qquad = 57.6$$

$$0.04 = k$$

The resistance of 90 inches of the same kind of wire with a $\dfrac{1}{4}$-inch radius is 57.6 ohms.

Step 5   Check.

We will check the equation $R = \dfrac{kl}{r^2}$ with $k = 0.04$ for $R = 12$, $l = 75$, and $r = \dfrac{1}{2}$ and for $R = 57.6$, $l = 90$, and $r = \dfrac{1}{4}$. The statement $12 = \dfrac{0.04(75)}{\left(\dfrac{1}{2}\right)^2}$ is true, and the statement $57.6 = \dfrac{0.04(90)}{\left(\dfrac{1}{4}\right)^2}$ is true. It checks.

PROBLEM SET 9

Write a formula for each of the following statements. Use k as the constant of variation.

1. y varies directly with the square of x.

2. A varies directly with the cube of t.

3. R is directly proportional to the square root of w.

4. p is directly proportional to the cube root of s.

5. C varies jointly with r and the square of t.

6. d varies jointly with v and t.

7. G is jointly proportional to m and the cube of t.

8. L is jointly proportional to w and the square root of d.

9. y is inversely proportional to the square of x.

10. y is inversely proportional to the cube root of t.

11. B varies inversely with the cube of a.

12. E varies inversely with the square of g.

13. C varies directly with a and inversely with b.

14. z varies jointly with x and y and inversely with w.

15. P is directly proportional to x and inversely proportional to the square of y.

16. F is jointly proportional to g and h and inversely proportional to m.

Solve the following variation problems using the five-step strategy.

17. y varies directly with the square of t. If $y = 24$ when $t = 4$, find y when $t = 5$.

18. P varies directly with the cube of t. If $P = 144$ when $t = 6$, find P when $t = 9$.

19. T is directly proportional to the square root of r. If $T = 3$ when $r = 16$, find T when $r = 25$.

20. L is directly proportional to the cube root of x. If $L = 12$ when $x = 27$, find L when $x = 64$.

21. z varies jointly with x and y. If $z = 20$ when $x = 6$ and $y = 5$, find z when $x = 7$ and $y = 9$.

22. E varies jointly with t and the square of v. If $E = 22$ when $t = 11$ and $v = 2$, find E when $t = 4$ and $v = 5$.

23. P is jointly proportional to x and the cube of y. If $P = 141.75$ when $x = 7$ and $y = 3$, find P when $x = 27$ and $y = 2$.

24. L is jointly proportional to s and t. If $L = 16.8$ when $s = 14$ and $t = 3$, find L when $s = 28$ and $t = 4$.

25. F varies inversely with the square of m. If $F = 15$ when $m = 3$, find F when $m = 5$.

26. T varies inversely with the square root of r. If $T = 29$ when $r = 9$, find T when $r = 25$.

27. S is inversely proportional to the cube root of t. If $S = 42$ when $t = 8$, find S when $t = 64$.

28. z is inversely proportional to the cube of x. If $z = 175$ when $x = 2$, find z when $x = 5$.

29. P varies directly with x and inversely with y. If $P = 170$ when $x = 5$ and $y = 3$, find P when $x = 7$ and $y = 4$.

30. F is jointly proportional to g and h and inversely proportional to the square of n. If $F = 45$ when $g = 81$, $h = 5$, and $n = 6$, find F when $g = 9$, $h = 12$, and $n = 3$.

31. Art's wages are directly proportional to the number of hours he works per week. If Art works 36 hours in a week, he earns \$540. What are his wages if he works 40 hours in a week?

32. When a gas is compressed at a constant temperature, the pressure of the gas varies inversely with its volume. If the pressure of a gas is 40 pounds per square inch when its volume is 28 cubic feet, find the pressure when its volume is 50 cubic feet.

33. The number of ounces in a cylindrical can is jointly proportional to its height and the square of the radius of its base. If a can with a radius of 4 inches and a height of 8 inches holds 10 ounces, find the approximate number of ounces in a can that is 10 inches high with a radius of 3 inches.

34. The weight of an object on the moon is directly proportional to its weight on Earth. A man who weighs 200 pounds on Earth weighs 32 pounds on the moon. How much would a 160-pound man weigh on the moon?

35. The revenue earned by a business is directly proportional to the number of items it sells. If the business sells 4250 items, its revenue is \$103,700. Find its revenue if it sells 5340 items.

36. The electrical resistance of a wire varies directly with its length and inversely with the square of its radius. If the resistance of 100 centimeters of wire with a radius of $\frac{1}{2}$ centimeter is 30 ohms, find the resistance of 75 centimeters of the same wire with a $\frac{1}{4}$-centimeter radius.

37. The distance an object falls varies directly with the square of the time it is falling. If an object falls 144 feet in 3 seconds, how far will it fall in 7 seconds?

38. A piece of land is being subdivided. The cost of a plot of this land is jointly proportional to its length and width. If the cost of a plot of land 100 feet by 90 feet is $50,000, find the cost of a plot of land that is 120 feet by 100 feet.

39. The intensity of light received by a light source varies inversely with the square of the distance from the light source. The intensity from a source is 9 foot-candles at a distance of 4 feet. What is the intensity when the distance from the light source is 10 feet?

40. The volume of a gas varies directly with its temperature and inversely with its pressure. If the volume of a gas is 15 cubic meters when the temperature is 125 Kelvin and the pressure is 25 kilograms per square meter, find the volume when the temperature is 150 Kelvin and the pressure is 20 kilograms per square meter.

10

Cumulative Review

The problems in this chapter are from Chapters 1–9, in no particular order. First determine which type of problem each one is and then solve it. Be sure to follow the five-step strategy and draw the appropriate chart or diagram for each problem.

Strategy for Solving Word Problems

1. <u>Read</u> the problem <u>and write</u> what is <u>given and</u> what you are asked to find (<u>the unknown</u>). It is also helpful to decide what type of problem it is.
2. <u>Begin</u> the solution <u>with</u> "<u>Let x = the unknown</u>." Represent other unknowns in terms of x. If possible, make a diagram or chart that relates the given and unknown quantities.
3. Go back, <u>reread</u> the problem, <u>and write an equation</u> that relates the given quantities.
4. <u>Solve the equation</u> and write your answer in a sentence using appropriate labels.
5. <u>Check</u> your answer back into the original words of the problem.

PROBLEM SET 10

1. Tickets for the Lion's Club Pancake Breakfast cost $4 for adults and $3.50 for children. If a total of 75 tickets were sold and $280 was collected, how many adults and how many children attended?

2. A chemist has one solution that is 40% acid and another that is 10% acid. How much of each should he mix to obtain 60 milliliters of a solution that is 20% acid?

3. The larger of two integers is two less than three times the smaller integer. If their sum is 18, find the two integers.

4. The hypotenuse of a right triangle is one more than the second side. The shortest side is 7 feet. Find the two missing sides.

5. Sam invested $20,000 in stocks and bonds. The stocks paid a 3% annual return, and the bonds paid a 5% annual return. If the return on the stocks was $40 more than the return on the bonds, find the amount he invested in each.

6. How long will it take $100 to double if it is invested at 5% simple interest?

7. The length of a rectangle is one more than twice its width. The perimeter is 65 centimeters. Find the dimensions of the rectangle.

8. The first side of a triangle is three times the second side. The third side is five more than the second side. If the perimeter of the triangle is 35 inches, find the lengths of its three sides.

9. Pat can fly his airplane 1455 miles with the wind in the same time it takes him to fly 1395 miles against the wind. If the speed of the wind is 10 miles per hour, find the speed of the airplane in still air.

10. In triangle ABC, the measure of angle A is equal to the measure of angle C. The measure of angle B is one-half the measure of angle A. Find the measures of the three angles.

11. How many ounces of pure boric acid should be added to 16 ounces of a 10% boric acid solution to obtain a 50% boric acid solution?

12. A bicyclist travels from her home to the beach and back in a total of 2.5 hours. Her rate going is 12 miles per hour, and her rate returning is 8 miles per hour. How far does the bicyclist live from the beach?

13. Five more than twice an integer is equal to six less than three times the integer. Find the integer.

14. Two trains leave Chicago, one headed due east at 55 miles per hour and the other headed due west at 65 miles per hour. In how many hours will the trains be 300 miles apart?

15. Antonio takes 6 hours to paint a fence alone. Erik can paint the same fence in 8 hours. How long would it take them to paint the fence together?

16. Jessica borrows some money at 8% simple interest. After 2 years she owes $568 in interest. How much did she borrow?

17. There were 125 people who attended the Mustang basketball game. Tickets for students sold for $5 each, and tickets for nonstudents sold for $8 each. If the total ticket sales for the game were $736, how many students attended the game?

18. y varies directly with the square of t. If $y = 54$ when $t = 3$, find y when $t = 5$.

19. A sailboat leaves the harbor traveling at 15 miles per hour. Four hours later a motorboat begins to pursue the sailboat at a speed of 55 miles per hour. How long will it take the motorboat to catch up to the sailboat?

20. How many pounds of peanuts that sell for $3 per pound should be mixed with 4 pounds of raisins that sell for $5 per pound to obtain a mixture that will sell for $3.50 per pound?

21. Samantha and Julie can weed their garden together in 4 hours. If Julie works alone, she can weed the garden in 6 hours. How long would it take Samantha to weed the garden alone?

22. A manufacturer produces items at a daily cost of 75 cents each. Its daily fixed costs are $300. If the company sells each item for $1, how many items does the company have to produce and sell each day to break even?

23. How much fuel that sells for $1.25 per gallon should be mixed with 200 gallons of fuel that sells for $2 per gallon to obtain a mixture that sells for $1.55 per gallon?

24. Three more than twice the reciprocal of a number is equal to $\dfrac{7}{2}$. Find the number.

25. The cash drawer at Shop-Rite Market contains $219 in bills. There are twice as many $5 bills as $1 bills and three more $10 bills than $1 bills. Find the number of each kind of bill in the drawer.

26. Dean can wash the family's car in 30 minutes, and his brother, Scott, can wash the car in 20 minutes. How long will it take them to wash the car together?

27. Angle A and angle B are complementary angles. The measure of angle A is three more than twice the measure of angle B. Find the measures of the two angles.

28. How much water should be added to 8 gallons of a 10% soap solution to obtain a 4% solution?

29. Jose has a total of 11 quarters and nickels in his pockets. The value of the coins is $1.95. How many of each coin does he have?

30. A 27-foot log is cut into two pieces. The longer piece is twice the shorter piece. Find the length of each piece.

31. Two cars left an intersection at the same time, one headed due north at 40 miles per hour and the other headed due west at 30 miles per hour. How far apart were the cars after 2 hours?

32. Mr. Gordon finds that he needs $12,000 in supplementary income. He has $150,000 to invest either in AA bonds that pay 10% annual interest or in a savings certificate that pays 5% interest per year. How much should he invest in each to earn the $12,000 per year?

33. The sum of three consecutive even integers is 42. Find the integers.

The following problems require techniques from intermediate algebra to solve.

34. A rectangle has a length that is 2 meters more than its width. If 1 meter is cut from the length and 1 meter is added to the width, the resulting figure has an area of 121 square meters. Find the dimensions of the original rectangle.

35. The owner of a $20,000 IRA decides not to add any more money to this account. If the account pays 5% annual interest compounded monthly, how much will be in this account in 20 years when the owner retires?

36. Joan's pool is 30 feet by 40 feet. She wants to spread redwood bark in a strip of uniform width around the pool. She has enough bark for 296 square feet. How wide should the strip be?

37. It takes two painters 2 hours to paint a room together. If each worked alone, the faster painter could do the job in 3 hours less than the other painter. How long would it take the faster painter to complete the job alone?

38. The intensity of light received by a light source varies inversely with the square of the distance from the light source. The intensity from a source is 8 foot-candles at a distance of 2 feet. What is the intensity when the distance from the light source is 5 feet?

Chapter 1

1. $x + 4$

2. $8 + x$ or $x + 8$

3. $5 + x$ or $x + 5$

4. $7 + x$ or $x + 7$

5. $6 + x$ or $x + 6$

6. $x + 8$

7. $9 + x$ or $x + 9$

8. $2 + x$ or $x + 2$

9. $x - 3$

10. $x - 5$

11. $8 - x$

12. $12 - x$

13. $x - 6$

14. $11 - x$

15. $x - 13$

16. $20 - x$

17. $3x$

18. $10x$

19. $6x$

20. $12x$

21. $0.50x$

22. $0.20x$

23. $\dfrac{3}{4}x$

24. $\dfrac{2}{3}x$

25. $\dfrac{4}{x}$

26. $\dfrac{x}{7}$

27. $\dfrac{10}{x}$

28. $\dfrac{x}{8}$

29. $\dfrac{x}{5}$

30. $\dfrac{20}{x}$

31. $2x + 3$ or $3 + 2x$

32. $10x + 6$ or $6 + 10x$

33. $3x - 2$

34. $6x - 5$

35. $\dfrac{1}{2x}$

36. $\dfrac{1}{x + 2}$ or $\dfrac{1}{2 + x}$

37. $2(x + 3)$

38. $3(x + 4)$ or $3(4 + x)$

39. $10(x - 12)$

40. $5(8 - x)$

For Problems 41–52, answers will vary. In each problem, let x = the number.

41. The sum of a number and six

42. Eight more than a number

43. The difference of 12 and a number

44. Ten less than a number

45. Five more than twice a number

46. Nine more than three times a number

47. Three less than four times a number

48. Two less than five times a number

49. Four times the difference of three times a number and six

50. Five times the sum of eight times a number and three

51. The quotient of a number and nine

52. The reciprocal of three less than a number

For Problems 53–62 and 65–68, let x = the number.

53. $x + 7 = 32$

54. $x + 12 = 102$

55. $x - 3 = 22$

56. $8 - x = 41$

57. $\frac{2}{3}x + 7 = 17$

58. $\frac{3}{4}x - 5 = 4$

59. $x - 3 = 19$

60. $x - 13 = 52$

61. $2x + 4 = 34$

62. $3x + 5 = 26$

63. Let x = first consecutive even integer; $x + x + 2 = 72$

64. Let x = first consecutive odd integer; $x + x + 2 = 82$

65. $2 + 3\left(\frac{1}{x}\right) = \frac{7}{3}$

66. $5\left(\frac{1}{x}\right) - 3 = -\frac{8}{3}$

67. $0.07x = 98$

68. $0.12x = 168$

Chapter 2

1. 21 and 28

2. 14 and 29

3. 17 and 53

4. 15 and 59

5. 15

6. 12

7. 7

8. 20

9. 21, 23, and 25

10. 11, 12, and 13

11. 29, 31, and 33

12. 18, 20, and 22

13. 7 ft and 17 ft

14. 14 in., 18 in., and 28 in.

15. Susan stuffs 30, Valerie stuffs 45, and Judy stuffs 15.

16. 14 cm and 31 cm

17. 17 and 28

18. 21 and 45

19. 13 and 20

20. 38 and 43

21. 12

22. 16

23. 32

24. 30

25. 3

26. 25

27. 12

28. 9

29. 10,456 women

30. Oregon: 1326; Washington: 2345

31. 5 and 15

32. 9 and 12

33. 6 and 12

34. 10 and 12

35. 7 and 9 or -7 and -9

36. 8 and 9 or -8 and -9

37. Gannon: 4689; Bledson: 4359

Chapter 3

1. 9 in. by 14 in.

2. 17 cm by 37 cm

3. 18 ft by 8 ft

4. 17 yd by 28 yd

5. 46 in., 12 in., and 15 in.

6. 26 ft, 7 ft, and 21 ft

7. 5 ft, 7 ft, and 9 ft

8. 9.5 ft, 14.5 ft, and 19 ft

9. 60°, 30°, and 90°

10. 25°, 53°, and 102°

11. 70°, 35°, and 75°

12. 37°, 33°, and 110°

13. 134° and 46°

14. 143° and 37°

15. 63° and 27°

16. 68° and 22°

17. 6.75 in.

18. 30.25 cm

19. 4.0 cm

20. 12.7 mm

21. 65 yd by 110 yd

22. 13 mi

23. 7.1 ft

24. 25 cm

25. 9 ft

26. 15 in.

27. 6 in.

28. 4 cm

29. 7 in.

30. 7 m by 15 m

31. 8 yd by 13 yd

32. 72 yd by 116 yd

33. 42.4 yd

Chapter 4

1. 19 dimes and 7 quarters

2. 5 dimes and 17 quarters

3. 15 nickels and 9 quarters

4. 20 nickels and 7 quarters

5. 13 dimes

6. 9 dimes

7. 15 first-class and 7 postcard stamps

8. 42 first-class and 18 postcard stamps

9. 8 45-cent and 12 20-cent stamps

10. 14 45-cent and 8 20-cent stamps

11. 68 adults and 25 children

12. 215 adults and 305 children

13. 84 students

14. 187 adults and 63 children

15. 53 adults

16. 52 adults

17. 4 $1 bills, 17 $5 bills, and 5 $10 bills

18. 6 $1 bills, 4 $5 bills, and 8 $10 bills

19. 3 dimes

20. 7 $5 bills

21. 5 nickels, 12 dimes, and 15 quarters

22. 64 children and 32 adults

23. $46

24. 8 dimes and 11 quarters

25. 17 dimes, 17 nickels, and 8 quarters

26. 9 paperback and 5 hardcover books

27. 11 singles and 20 doubles

28. 37 dimes and 42 quarters

29. 83 students and 29 nonstudents

30. 51 adults and 34 children

31. 50 first-class and 8 postcard stamps

Chapter 5

1. 20 liters

2. 4 lb

3. 15 gal of 10%, 25 gal of 50%

4. 13 oz of 4%, 26 oz of 1%

5. 8 qt

6. 20 lb

7. 9 oz of 10%, 15 oz of 50%

8. 180 oz of 5%, 120 oz of 10%

9. 8 oz of 35%, 20 oz of water

10. 12 oz of water, 24 oz of 12%

11. 6 qt

12. 48 cc

13. 20 oz

14. 15 gal

15. 20 cc

16. 4 cc

17. 84 ml of 10%, 24 ml of pure

18. 26 qt of 10%, 10 qt of pure

19. 4 oz

20. 2.5 qt

21. 4 lb of jelly beans,
8 lb of chocolate eggs

22. 9 lb of raisins, 18 lb of nuts

23. 8 lb of Kentucky, 6 lb of rye

24. 5 gal

25. 5 gal

26. 10 lb of mixed nuts, 5 lb of M & M's

27. 6 oz of 60%, 9 oz of 10%

28. 8 oz of 25%, 12 oz of 60%

29. 1 gal

30. 6 lb of peanuts, 3 lb of raisins

Chapter 6

1. $4350 in stocks, $5650 in bonds

2. $900 in checking, $1600 in savings

3. $360 in bonds, $640 in savings

4. $1700 in stocks, $3300 in bonds

5. $6000 at 5%, $4000 at 9%

6. $15,000 at 4%, $25,000 at 6%

7. $32,000 at 6%, $18,000 at 8%

8. $8000 in bonds, $12,000 in stocks

9. $10,000

10. $37,500

11. 40 yr

12. 20 yr

13. $550 from father, $450 from uncle

14. $840 at 4%, $420 at 2%

15. $210 at 2%, $630 at 5%

16. $4500 at 3%, $1500 at 5%

17. $5500 at 4%, $2500 at 6%

18. $1400 in stocks, $3600 in bonds

19. $8200 in CD, $1800 in stock

20. $3300 in bonds, $1700 in stocks

21. 103 pairs of sunglasses

22. 510 keychain–bottle openers

23. 43 watches

24. 83 travel clocks

25. $2734.52

26. $3536.18

27. $635.37

28. $2141.00

Chapter 7

1. 120 mph; 230 mph

2. 2.5 hr

3. 4 mph; 5 mph

4. 6 hr; 360 mi

5. 3 hr

6. 10 hr

7. 5 hr

8. 1.2 hr

9. 24 mi

10. Jeff—24 mph; Jim—71 mph

11. 7.4 hr

12. Judith—4 mph; Danya—9 mph

13. 6 mph

14. 500 kph

15. Current—3 kph; in still water—17 kph

16. Wind—2 mph; in still air—12 mph

17. Pursuit; 10 hr

18. Round-trip; 180 mi

19. Opposite direction; 3.5 hr

20. Same direction; Rachel—5 mph; Ann—4 mph

21. Current; Wind—1.5 mph; in still air—7.5 mph

22. Same direction; 5 hr

23. Opposite direction; 1.5 hr

24. Current; 8 mph

25. Round-trip; $\frac{1}{4}$ hr or 15 min

Chapter 8

1. 6 hr

2. $3\frac{3}{7}$ hr

3. $1\frac{1}{5}$ days

4. 18 min

5. 12 min

6. $1\frac{13}{47}$ days

7. $37\frac{1}{2}$ hr

8. $13\frac{1}{3}$ hr

9. 6 hr

10. $7\frac{1}{2}$ hr

11. 15 hr

12. 30 hr

13. 12 sec

14. $2\frac{1}{17}$ hr

15. 18 hr

16. 60 minutes or 1 hr

17. $\frac{6}{7}$ hr

18. $25\frac{5}{7}$ min

19. 8 min

20. $1\frac{1}{8}$ hr

21. $3\frac{2}{3}$ min

22. $9\frac{2}{3}$ hr

23. $7\frac{7}{10}$ hr

24. $1\frac{5}{6}$ hr

Chapter 9

1. $y = kx^2$

2. $A = kt^3$

3. $R = k\sqrt{w}$

4. $p = k\sqrt[3]{s}$

5. $C = krt^2$

6. $d = kvt$

7. $G = kmt^3$

8. $L = kw\sqrt{d}$

9. $y = \dfrac{k}{x^2}$

10. $y = \dfrac{k}{\sqrt[3]{t}}$

11. $B = \dfrac{k}{a^3}$

12. $E = \dfrac{k}{g^2}$

13. $C = \dfrac{ka}{b}$

14. $z = \dfrac{kxy}{w}$

15. $P = \dfrac{kx}{y^2}$

16. $F = \dfrac{kgh}{m}$

17. 37.5

18. 486

19. 3.75

20. 16

21. 42

22. 50

23. 162

24. 44.8

25. 5.4

26. 17.4

27. 21

28. 11.2

29. 178.5

30. 48

31. $600

32. 22.4 lb/in.^{2}

33. 7 oz

34. 25.6 lb

35. $130,296

36. 90 ohms

37. 784 ft

38. $66,666.67

39. 1.44 foot-candles

40. 22.5 m^3

Chapter 10

1. Coin; 35 adults and 40 children

2. Mixture; 20 ml of 40% and 40 ml of 10%

3. Integer; 5 and 13

4. Geometry; 24 ft and 25 ft

5. Finance; $13,000 in stocks and $7000 in bonds

6. Finance; 20 yr

7. Geometry; 10.5 cm by 22 cm

8. Geometry; 6 in., 11 in., and 18 in.

9. Motion; 475 mph

10. Geometry; 72°, 36°, and 72°

11. Mixture; 12.8 oz

12. Motion; 12 mi

13. Integer; 11

14. Motion; 2.5 hr

15. Work; $3\dfrac{3}{7}$ hr

16. Finance; $3550

17. Coin; 88 students and 37 nonstudents

18. Variation; 150

19. Motion; 1.5 hr

20. Mixture; 12 lb

21. Work; 12 hr

22. Finance; 1200 items

23. Mixture; 300 gal

24. Integer; 4

25. Coin; 9 $1, 18 $5, and 12 $10 bills

26. Work; 12 min

27. Geometry; 61°, 29°

28. Mixture; 12 gal

29. Coin; 7 quarters and 4 nickels

30. Integer; 9 ft and 18 ft

31. Motion; 100 mi

32. Finance; $60,000 at 5% and $90,000 at 10%

33. Integer; 12, 14, 16

34. Geometry; 10 m by 12 m

35. Finance; $54,252.81

36. Geometry; 2 ft

37. Work; 3 hr

38. Variation; 1.28 foot-candles

Judy Barclay received her bachelor's degree from the State University of New York College at Cortland and her master's degree from the University of Massachusetts. She has been teaching for thirty years at the secondary and community college levels. She is currently the Mathematics Division chair at Cuesta College in San Luis Obispo, California. She has taught both elementary and intermediate algebra many times and also teaches a short course devoted to helping students learn to solve word problems.

NOTES